Machine Learning for Marketing

Hiroshi Mamitsuka

International Standard Book Number (ISBN): 978–4–9910445–2–6 (Printed book), 978–4–9910445–3–3 (eBook)

Library of Congress Control Number:2019906946

To Miko

and

Hiro

Preface

I began doing research on machine learning when I started working with a company's research laboratory. It was more than 25 years ago. The company believed in that machine learning would be helpful for the company business in the future, and the research group of only a few people started working on machine learning research in the company, just one year before I joined the company. I'm not working with that company any more, while the company's expectation would have been correct. Machine learning, currently a major part of artificial intelligence (AI), is not only useful for development of science and technology but also a key driving force for advancing our society. In reality, our life has changed or is changing by many AI-based autonomous technologies, including robotics, security-alert systems, face recognition, self-driving, etc. Eventually our life will be redefined by the advancement of AI in the near future. The research laboratory where I joined more than 25 years ago has a machine learning research group still now, while the big difference from 25 years ago is that the current research group has more than 100 people.

At that time my work was not doing research only. I visited customer companies to talk over their data and requests on what they want to know from the data. Also my visit of customer companies was not only just for talking and meeting, but also for doing more substantial work, like running machine learning software (developed by my group) over the data owned by customers. The customer data were confidential and I was unable to carry the data out of customer companies, and then I had to stay at each customer company on the whole day, or commute there all a week or entire a month. As such I tackled many problems, which customer companies had and provided. Of course they were not always issues in marketing, but some of them were definitely on marketing. A typical example is customer churn analysis, campaign management to increase loyal customers, etc. I worked on the churn problem with many customers, which include internet service providers, mobile phone carriers, E-commerce (electric commerce) site developers, etc. Although I tackled such problems in marketing, honestly I had no good knowledge on marketing and more importantly no systematic understanding of marketing. In other words, I had no comprehensive understanding on what and how I could contribute to the customer companies or more generally the society, though practically problems I tackled could be solved by machine learning software we provided.

Six years ago, I had a chance to work with a sort of companies in one project, which builds a database to be shared by them and generates software to retrieve useful knowledge from the database. I worked as a consultant in this project, which reminds me a lot of past experiences when I was working with a company and also customers. First I noticed my knowledge on business was definitely limited, which made me, being involved with this project, start reading a lot of textbooks, mainly used in regular courses of Master of Business Administration (MBA). Reading these books, which are sometimes on marketing, I noticed that a wide variety of aspects of marketing can be changed to be done in a more systematic manner by using "data". In other words, machine learning would be useful for a lot of points of marketing to which machine learning or AI has not been considered to apply yet. In fact, each step of target marketing, i.e. segmentation, targeting and positioning, can be autonomous by using AI, if marketing strategies are properly formulated into mathematical problems and also good amount of high-quality data are given. For example, an important objective of marketers in target marketing can be interpreted as a problem of finding the market segments which have a good number of customers but have no powerful competitors. This objective can be formulated into a machine learning problem, in which parameters can be optimized from data. This is exactly the work of machine learning and then autonomous AI. Relationship marketing would be more so, because customer purchase records are more abundant and also accessible. These data are definitely useful to solve machine learning problems set for relationship marketing by optimizing model parameters. In short machine learning over such data would be able to show what marketers want to see or directly perform what marketers want to do. This would be reasonable, while I then checked a sort of academic journals on marketing and related fields and noticed that machine learning has not been used for marketing yet extensively or we are just standing at the entrance of the world of autonomous marketing by machine learning or artificial intelligence.

This book covers many aspects and problems derived from marketing, particularly rather traditional or standard marketing, for which machine learning can be applied to solve. Some of them would be already well known, while I think most of them are not necessarily so. I would be happy if this book contributes to using machine learning more widely for marketing than now.

Hiroshi Mamitsuka
Uji, Japan / Espoo, Finland
June, 2019.

Contents

Chapter 1

Introduction

Various types of data have been stored in a lot of fields, such as science, engineering, economics, agriculture, finance, etc., and so we are in the era of *big data*. The scale advantage of abundant data should be used effectively, to make our life more comfortable and richer, for example, more advanced sciences and technologies, more affordable healthcare, higher efficiency of business activities, and so on. One example is marketing in business, where a variety of data on products, customers, services, companies and their interactions, like user-item purchase records and social relationships between customers, have been already obtained and further increasing [23]. Numerous aspects of marketing can be application targets of so-called *data-driven* approaches, such as business analytics, data science, data mining and machine learning, which lead to predictive *marketing analytics* [111, 16], or more generally artificial intelligence [46, 35, 97]. Typically, recommendation has already long history, being started with rather a simple idea of, for a target customer, finding similar customers with previous purchase records and suggesting the target customer items which are not yet purchased but bought by most of similar customers. This intuitive concept, called *collaborative filtering* [47], behind recommendation would be reasonable, consistent with the thoughts owned by marketers, for a long time. Customer churn analysis would be another example, which is more suitable to be addressed by machine learning to predict current membership customers to potentially cancel the contract in the near future, by using past and present customer data and records [2]. More concretely, customer churn prediction is a proper application of binary classification, providing machine learning research with an interesting, challenging problem of class imbalanceness on training data due to a limited number of past (but not current) customers. These two application examples are just a part of iceberg above water, and other applications have been addressed and at the same time, numerous opportunities would be still waiting for data-driven approaches to be applied in marketing. In other words, the current applications including the above two examples might be rather exceptional, since they have been already well considered and developed by data-driven techniques, particularly machine learning, but we think that a plenty of promising situations to be solved by machine learning still remain unsolved, untouched or even unconceived in marketing.

However recent issues in business magazines and academic journals on marketing and consumer research show that machine learning-based work for new applications are extensively emerging in marketing, particularly using the data, especially those called *user-generated content* (UGC) from social networks or social media as the main input. A short list of examples of the new applications using images and text of social networks (or social media) includes retrieving online social ties [114], extracting brand information [51, 66, 55, 110, 49], understanding consumer needs [101], consumer sharing [108] and consumer behaviors on advertising [68], detecting key terms relevant to sentiment patterns [109], measuring social media influencer index [6], to name just a few. Also online pricing is definitely an interesting issue in marketing [93, 74]. In fact these work mostly have appeared in the last few years, and before that it was rare to meet such papers. We can say that promising applications now appear here and there in marketing, making use of the predictive power of machine learning. A noteworthy feature of these work is that the used data are obtained through modern information sources, i.e. internet, such as social media or social networks, and the applications are motivated by those data. Although however first they look totally new as applications, we can then notice that these applications can also use the traditional data, such as survey data by customers over products which eventually has the same format and information as the data obtained from reviews in E-commerce sites or blogs. In other words, currently marketing researchers apply machine learning to the modern data directly, skipping doing so over the traditional data, though the same applications can be made from the data in traditional marketing, such as target marketing and relationship marketing. This indicates that it would be important to consider to use machine learning over the traditional framework of marketing, particularly over the data already obtained and accumulated. Another characteristic point of note on machine learning in marketing currently is that text mining is well used to capture knowledge (useful for marketing) from particularly online text data, so-called *electric word-of-mouth* (eWOM), including blogs and reviews. In fact, a lot of work of machine learning in marketing, including those shown in the above, are in this line of research [49, 34, 78, 96, 24]. Text mining in marketing however would be rather more natural language processing than machine learning, and so will not be touched in this book.

Machine learning can be used for not only analyzing the stored data to understand the past results but also for predicting the future of the changing world of marketing [57]. Also as mentioned in the above briefly, machine learning can be applied to not only recent online data but also traditionally obtained data in marketing. We can then say that now the two sides, i.e. machine learners and marketers, may understand the other side more to explore and solve various new problems/applications in marketing by machine learning. A key point is to build a proper machine learning approach for each application. This stage would need a proper guide, which presents ways of defining machine learning problems and developing methods for solving the problems from data given in marketing, particularly not necessarily only new data like social networks and related data but also traditional data like customer demographic data, user survey, purchase records and user-item matrix for recommendation. In fact machine learners can design, for a given problem, a model and its algorithm flexibly, according to given data or the assumption behind the problem (even under the same problem setting), instead of just setting up a problem to which existing methods can be applied. This manner of understanding machine learning would be key for those who, in the application side, are likely to apply existing machine learning approaches straightforwardly to each problem setting. In fact again as mentioned above, although text mining has been well used in marketing for understanding eWOM, text mining would be also more likely to use existing machine learning methods as tools rather directly, e.g. text classification in marketing [40], tagging online contents automatically [85], and also text analysis for consumer research [48]. Thus again text mining will not be mentioned in this book.

There already exist books on the applications of machine learning, data mining or rather data science (business analytics) to marketing [28, 64]. Their contribution to marketers, particularly practitioners working on real data, would be tremendous. For example, [28] provides a comprehensive, practical, step-by-step guide of applying data science techniques to market segmentation, showing even codes in R. The authors of these books, basically from the application side, assume that machine learning methods are already given, and then each problem is set so that machine learning algorithm can be easily applied. This would be reasonable. On the other hand, however, an important point would be, again, to show how machine learning methods can be (uniquely) designed for a given application, and can be changed, according to available data or the assumption behind the application. Over all we can again emphasize that machine learning models can be designed by the idea/ assumption behind the problem setting.

The objective of this book is primarily to show the most appropriate (and simplest) machine learning method for the given problem and its assumption, and then present a way of exploring proper changes of machine learning methods, according to the variation of data, problem settings and/or assumptions behind even a single problem setting. The machine learning methods in this book are kept as simple as possible, particularly comparing with the recent, more complicated, black-box-type methods, such as deep neural networks. We think that by using such simple models, readers will understand each model and its variations more, particularly reasonably the model being modified, according to given data and the assumption behind the problem. We hope that this book will be helpful for marketers to understand machine learning models in the manner described (Interested readers can refer to machine learning books more, e.g. [70]), and for machine learners to understand the idea and points of applications in marketing.

In this book, we focus on two mainstream (and traditional) concepts of marketing: *target marketing* and *relationship marketing*, which have been established already and now well matured in marketing. In fact both are key ideas generated and developed in the modern marketing history. Additionally we consider database marketing, which would be applications more suitable for machine learning. We explore the possibility of developing a machine learning method for each step (or aspect) of these two concepts, assuming that we have a plenty of related data in marketing, such as those for customers, products, manufactures, their interactions, such as purchase records, and whatever.

The organization of this book can be given below.

Chapters 1 to 4 are the introductory part of this book, providing basic knowledge and information, which must be well understood for learning both machine learning and marketing.

Chapter 1, this chapter, purely introduces the ideas behind this book. Chapter 2 provides the concepts and terminologies in machine learning first and then marketing. Again this book focuses on two main concepts of marketing, i.e. target marketing and relationship marketing. Chapter 3 introduces the fundamentals of target marketing, which consists of three steps, called the STP (segmentation, targeting and positioning) strategy. These three steps are described in detail but concisely. Also the STP strategy in general uses several interesting ideas, such as SWOT (strengths, weaknesses, opportunities and threats) analysis and perceptual map. This chapter describes these ideas briefly but clearly by using several examples for each of them. Chapter 4 describes the basics of relationship marketing. In particular, this chapter focuses on several topics, such as RFM (recency, frequency and monetary) analysis, customer loyalty satisfaction, retention marketing (marketing funnel), which are described with some examples.

Chapters 5 to 8 are the main part of this book, presenting various machine learning approaches for marketing applications and their variants. The key point is we can change, for example, the data, problem setting and/or assumption behind the problem, and for each time, we can develop a machine learning method for each particular problem setting. We then describe such diverse models in these chapters, resulting in showing diverse machine learning approaches in these chapters.

Chapter 5 starts with fundamental machine learning concepts and approaches, which are required to address the problems in marketing to be introduced in later chapters. The machine learning paradigms in this chapter can be first classified into three concepts: regular machine learning, feature learning and kernel learning. In particular, regular machine learning focuses on two data types: vectors and nodes in a graph, and then for each type, three machine learning concepts, i.e. supervised, unsupervised and semisupervised learning, and so methods for totally six combinations (two data types × three learning concepts) are described. Feature learning has two types, feature selection and feature generation (dimensionality reduction), which both are described concisely. Finally for kernel learning, a standard procedure to make regular machine learning procedure kernelized into the corresponding method in kernel learning.

In Chapter 6, we focus on target marketing or the STP (segmentation, targeting and positioning) strategy, and explore the possibility of developing machine learning methods or applying the ideas of machine learning, for many aspects of the STP strategy. For example, clustering has been applied to segmentation, while in this book, we consider a variety of machine learning approaches for segmentation, depending on the data and assumptions we can consider. Also we will think about machine learning-based segment evaluation and SWOT analysis. Furthermore we present machine learning methods for generating perceptual maps, depending on the assumptions behind the maps, which are used for positioning the products, brands and companies. Finally we consider an interesting problem setting of finding the area in a perceptual map (or any dimensional space) which have no competitive products but a lot of customers. We then present a simple machine learning model and algorithm to address this problem.

Chapter 7 focuses on relationship marketing, and explores the possibility of building ML models to apply each of all possible parts of relationship marketing. This chapter contributes to the three aspects of relationship marketing: customer relationship management (CRM), retention marketing and market communications. In CRM, we consider three problems: learning user-item (customer-product) matrix, detecting profitable customers (automatic RFM (recency, frequency, monetary) analysis) and customer churn analysis. Then for retention marketing, we set up a general setting of problems, considering flexible transitions between different stages in marketing funnel, instead of incremental or progressive promotion through the marketing funnel. For this setting, we present two types of ML solutions. Finally in this chapter, we address a problem of optimally selecting communication channels among a wide variety of choices on current market channels, assuming that a good amount of data can be given to optimize the built model.

Finally Chapter 8 focuses on settings, in which not only one but multiple matrices are given as input. For example, given two or more matrices which always share one dimension, e.g. rows for instances, and in which the other dimension can be changed, like that different feature sets can be given, we may then want analyze these matrices, for finding interactions between different feature sets of different matrices or capturing common factors shared by different feature sets, etc. In this chapter, under the setting of given multiple matrices, we consider four types of ML settings: 1) supervised, 2) unsupervised (clustering), 3) unsupervised (factorization) and 4) learning interactions between features. Also we change the setting of given multiple matrices under the above four machine learning settings, while our focus goes more on to matrix factorization eventually, so-called *collaborative matrix factorization* for given multiple matrices [115]. Overall in this chapter, we present numerous types of collaborative matrix factorization, considering general combinations with an arbitrary number of matrices, as a representative data integrative approach.

Chapter 2

Concepts and Terminology

2.1 Data Types

We first explain data types used in machine learning. In machine learning, a database has records, where each record is called an *instance*, an *example* or a *sample*. In this book we use the term "instance" for each record. Then the data type of each instance can be categorized into six types: a vector, a set, a sequence (or a string), a tree, a graph and a node in a graph. Below we will explain each type of data. Also we note that currently in real-world applications, each record can have more than one data type at the same time, such as a vector and a node in a graph. We will explain this type of data also after the description on the six data types.

2.1.1 Vector

When each instance has a certain number of values, which in machine learning, are called *features*, *attributes* or *variables*, and in this book we use *features* for values of an instance, where features can be numerical or categorical. Vectors are standard data in machine learning and data mining. Fig. 2.1 shows an illustrative example of vectors (called *feature vectors*) for instances, resulting in a matrix for multiple instances. Here are features, terminology and examples:

Building blocks: are feature values, which can be either discrete or continuous.

Example 1: demographic data: One instance is an individual, where features are age, gender, etc.

Example 2: gene expression: One instance is a gene, and features are experimental conditions, under which expression of the corresponding gene is measured. Each matrix element is the expression value of the corresponding gene and the corresponding experimental condition.

	Feature 1	Feature 2	Feature 3	Feature 4
Instance 1	A	-1.2	Large	0.5
Instance 2	B	1.8	Small	-0.4
Instance 3	B	-0.3	Medium	-0.2
Instance 4	A	1.9	Small	0.3
...				

Figure 2.1: Instances, each being a vector, are a matrix.

Instance 1: $\{A, B, C, D\}$, Instance 2: $\{D, C, B, A\}$

Figure 2.2: Two instances, each being a set.

2.1.2 Set

Each instance is a set, which has features (elements), where the number of features can be different for each instance and features have no order. Entirely a dataset can be a set of sets. Fig. 2.2 shows an illustrative example of sets.

Building blocks: One set has an arbitrary number of elements (features), which can be discrete or continuous. Also elements have no orders. Thus, for example, in Fig. 2.2, {A, B, C, D} and {D, C, B, A} are the same. We can say that a vector is a special case of a set.

Example: market basket: One typical example is a *market basket*, which is a set of items purchased by one customer at, for example, a department store, a convenience store or an E-commerce site. In general the number of items is always very large, while each user buys only a few number of items, which makes the market basket data very *sparse*.

2.1.3 Sequence and String

When elements in one set are ordered, the set becomes a *sequence*. In particular elements are a finite number of letters, such as the alphabet, the sequence is a *string* (see below in some more detail). As well as sets, the length of a sequence is flexible and not fixed.

Thus a sequence is a special case of a set, in the sense that the elements are ordered. Then a vector can be a further special case of a sequence, in the sense that the number of features is fixed. Fig. 2.3 shows an illustrative example of sequences. Below we describe the definition and terminology of sequences.

Discrete element for sequences: Discrete elements of sequences are called *letters* or *characters*. The set of letters is called the *alphabet*. A sequence of letters is called a string.

Instance 1: $DCBA$, Instance 2: $ADBCA$

Figure 2.3: Two instances, each being a sequence.

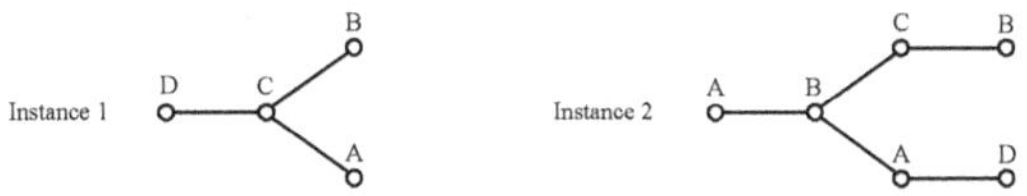

Figure 2.4: Two instances, each being a tree.

Subsequence and substring: A consecutive part of a sequence is called a *subsequence.* Also a consecutive part of a string is called a *substring.* For example, in Instance 1 of Fig. 2.3, DC and DCB are substrings, while DCA, i.e. the first, second and fourth letters, is not a substring.

Example 1: natural language: A typical sequence example is natural languages, which are the main data of computational linguistics, natural language processing and speech recognition, etc., which are classical applications of machine learning.

Example 2: gene/protein sequence: A nucleic acid sequence can be a string of four letters (corresponding to four types of nucleic acids). Similarly an amino acid sequence can be a string of twenty letters (corresponding to twenty amino acids). The length of one sequence/string can be different for both nucleic acid and amino acid sequences.

2.1.4 Tree

Each instance is a tree, and so the input data is a set of trees. Trees can be easily defined if graphs are already defined, and so suppose that graphs are already defined, trees are graphs without any cycling edges (or cycles).

Fig. 2.4 shows illustrative examples of trees, where Instance 1 has a branch from C to B and also to A, and similarly Instance 2 has a branch from B to C and A.

Building blocks: A tree consists of nodes and edges.

Cycles: Trees have no cycles. This discriminates trees from graphs.

Labels: A tree with labels is called a *labeled tree.* For example, in Fig. 2.4, A, B, C and D are labels. In this book, we consider labeled trees.

Root: In general, any node of a tree can be a *root,* while if one node of a tree is fixed as the root, the tree is called a *rooted tree.* For example, D of Instance 1 can be fixed as the root, and A of Instance 2 can be fixed as the root. We focus on rooted trees, and so we consider *labeled rooted trees.*

Leaves and internal nodes: In the rooted tree, except the root, we regard all nodes with only one edge as *leaves*. Nodes except leaves and the root are called *internal nodes*.

Ordered tree: For a rooted tree, nodes can be ordered from the root to leaves, and the tree is an *ordered tree*. Ordered trees are generated by allowing branches in sequences. In other words, sequences are a special case of ordered trees, generated by not allowing any branches.

Parent-children: In the rooted tree, for two nodes connected by an edge, the node closer to the root is called *parent* and the other node is called a *child*. Comparing with sequences, a feature of trees is a parent can have more than one children, while in sequences (and strings), a parent can have only one child.

Siblings, ancestors and descendants: Nodes with the same parent are called *siblings*. For a child, any node between its parent and the root are all *ancestors*. Similarly for a parent, its children and any node between the children and leaves are all *descendants*.

Depth: The number of edges from the root to a node is called the *depth* of the node. Usually the depth of the root is zero. The nodes with the same depth are called nodes at the same *layer*. On the other hand, the number of edges from a leaf is called the *height*. Usually the height of an leaf is zero.

Subtree: We explain regular definition of *subtree* below, while in frequent subtree mining, subtree means any connected part of a tree. Thus this definition in frequent subtree mining is different from the regular definition. In fact the regular definition is based on the definition of subsequence in sequences/strings, while the above definition (any part of a tree) is derived from the definition of subgraphs.

In regular definition, a subtree has some node (of the original tree) as the root (of this subtree) and also all nodes and edges of the leaf side of the root. For example, in Instance 2 of Fig. 2.4, if the root is the most left-hand side A, CB and AD of the most right-hand side can be subtrees, while BAD cannot be a subtree, because if B is specified as the root of a subtree, the subtree must be B(AD)CB.

Example: glycan (carbohydrate sugar chain): Building blocks of glycans are around 10 to 15 types of monosaccharides, which can be letters (labels). Glycan is generated by the connection (binding) of these monosaccharides, and the connection allows branches, resulting in trees with labels of monosaccharides.

2.1.5 Graph

One instance is a graph and input data are a set of graphs. Fig. 2.5 shows illustrative examples of graphs. Comparing with trees, one important feature of

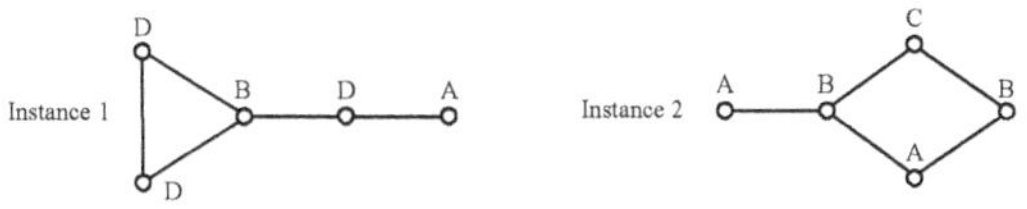

Figure 2.5: Two instances, each being a graph.

graphs is that graphs have cycles. For example, in Instance 1 of Fig. 2.5, two D and B are connected each other in the left-hand side, which turns into a cycle. This is not allowed in a tree, and trees are a special case of graphs.

Building blocks: A graph consists of nodes and edges, which connect nodes.

Degree: The number of edges connecting to one node is called the *degree* of the node.

Complete graph: If all nodes in a graph are connected each other, this graph is called a *complete graph.*

Labels: If nodes in a graph is labeled, this graph is called a *labeled graph.* Instances in Fig. 2.5 are two labeled graphs.

Direction of edges: If edges have some direction, the graph is called a *directed graph*; otherwise, the graph is called an *undirected graph.*

Routes: If one node is reached from another node by using one or more edges, the edges and nodes between the two nodes are entirely called a *route* from one of the two nodes to the other node.

Self-loop and cycles: If one edge connects the same node, this edge is called a *self-loop.* If two different nodes connected by one route are the same node, this route is called a *cycle.* The cycle allows to go from a node to another node by using more than one routes. Again graphs with no cycles are trees, meaning that a tree always has only one route between any two nodes.

Subgraph: A part of a graph is called a *subgraph.* If all nodes and edges are connected each other in a subgraph, the subgraphs is called a *connected subgraph.* We consider only connected subgraphs as subgraphs in this book.

Isomorphism: If two graphs have the same structure in terms of nodes and edge connectivity, they are called *isomorphic.*

Example: chemical structure of chemical compounds: The types of (physical) elements of chemical compounds are limited, such as carbons and oxygens, and they can be labels. Then the chemical structure (called a *molecular graph*) of a chemical compound can be regarded as a labeled graph.

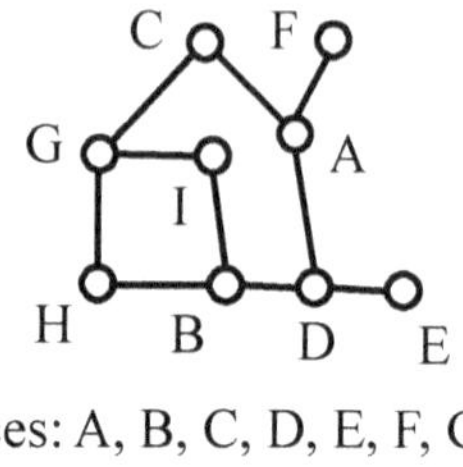

Figure 2.6: Instances are nodes in a graph, which is the entire data.

2.1.6 Node in a Graph

One instance is a node in a graph, and the entire data is a graph (all nodes in one graph is a set of all instances). Fig. 2.6 shows a simple, illustrative example. Nodes are instances, meaning that nodes are all unique (Note that this does not mean that labels of nodes are all different).

Example 1: social network: A social network is a graph with nodes for individuals and edges for some relationship between the individuals connected by the corresponding edges. Note that nodes (individuals or people) are all unique, and they can be assigned by labels, such as a male or a female.

Example 2: gene regulatory network: A gene regulatory network is a collection of molecular-level biological knowledge on gene regulations, such as that gene A is regulated by gene C. Regarding gene regulations as binary relations, i.e. two nodes connected by one edge, a graph with all binary relations of gene regulations is a gene regulatory network, where nodes are all unique genes. Again in this case also, genes can be with labels, such as gene functions.

2.1.7 Multiple Data Types

The last data type is that two or more data types can be given at the same time. For example, we can have a vector and a node in a graph both for each instance. We can just raise a couple of examples for this type of data.

Example 1: sequence and graph: drug-target interactions

An interaction of a drug and a target has both a drug and a target for one instance. A target is a protein, which can be represented by an amino acid sequence[1], while a drug is a small chemical compound, which can be represented by a chemical structure, corresponding to a molecular graph, i.e. a graph. Thus overall each instance of drug-target interactions can be represented by a pair of a sequence and a graph.

[1] If we think that a target is a gene, that can be represented by a nucleic acid sequence.

Example 2: vector and node in a graph: demographic data and social network

Demographic data consists of individuals with their features like age, sex, etc., meaning that each individual is a vector in demographic data. Also instead of demographic data, we can consider purchase data in E-commerce, which is generally a huge matrix of individuals (users) for rows and items (goods) for columns. Thus an instance can be each individual, which has features of purchase records, resulting in a vector for each individual. On the other hand, a social network is a graph with nodes of individuals, meaning that each instance is a node in a graph. Thus, entirely, given demographic data and social network means that each instance is a pair of a vector and a node in a graph.

Example 3: vector and node in a graph: gene expression data and gene network

Gene expression is typical vector data, in which each instance is a gene, which has expression values under multiple experimental conditions as features. On the other hand, we can have a network of genes from a lot of different biological aspects[2], such as gene regulation network, protein-protein interactions, signal processing network and metabolic pathways, etc. All these networks can be a graph with genes as nodes and some relationships between two genes can be an edge, resulting in that each instance is a node of a graph. Thus overall, one instance is a vector and also a node in a graph.

2.2 Machine Learning (ML)

We now briefly describe machine learning, focusing on introducing key terms to be used. Readers, particularly entry-level machine learners, are recommended to go through this chapter without trying to completely understand the terms within this chapter.

2.2.1 Paradigms

As mentioned earlier, data, input of machine learning, are a set of records, each being called an *instance*, *example* or *sample*. We use the "instance" through this book. Each instance has values, which are called*features*, *variables*, *attributes* or *variates*, where these features are visible and unchanged (during learning) and so called *observable variables*. We use "features" among the above four terms.

Occasionally instances can be classified into more than one categories, typically two (binary). Here are two examples:

Example 1: Mobile phone customers of some company can be segmented into at least the following two: *current subscribers* and *former subscribers* (though current subscribers might be categorized into more details, due to the risk

[2]Genes and proteins are synonyms, which can be alternatively used.

of leaving. Also promising subscribers among totally non-subscribers would be also a possible, important category for mobile phone companies).

Example 2: Regarding some disorder, *patients* and *healthy people* (also patients would be categorized into more, due to the seriousness of the disorder).

As such, when instances can be segmented into categories, each category is called a *class*, and each of the above *current subscribers*, *former subscribers*, *patients* and *healthy people*, etc. is called a *label*. The label is the name of a class, and if there are five labels, this means there exist five classes. Here are more examples:

Example 3: A customer is an instance, where the evaluation by one customer against some item is a label. Also another possible label is if a customer buys an item or not, which results in binary labels.

Example 4: A gene is an instance, where its function can be a label.

In Example 3, the evaluation can be binary or some moderate-size number, like a five-stage, while in Example 4, the functions of genes cannot be even by such a small number but a larger number. These two examples imply that labels would not be necessarily assigned to all instances (for example, functions are not necessarily assigned to all genes), implying that assigning labels would be a possible problem setting. More concretely predicting functions of unknown genes is a useful problem setting. This means that a label is a feature value to be predicted in machine learning.

Again all instances do not necessarily have labels, because obtaining labels is expensive. For example, in Example 3, we need to ask customers to give some evaluation against each item. Also we need some biological experiments to annotate functions to genes in life sciences, which needs huge cost and also time.

Thus machine learning has a paradigm which does neither assume classes nor labels. This setting is called *unsupervised learning*. The objective of unsupervised learning is to understand the data *distribution* or more generally to summarize data. A typical approach for summarizing data is *clustering*, where instances are grouped into several *clusters*. Clusters are different from observable variables like features, while variables corresponding to clusters must be trained from given data. Thus this variable corresponding to a cluster is called a *latent variable*. Estimating latent variables means clustering.

On the other hand, we call learning using labels *supervised learning*, where inputs are features of instances. The objective of supervised learning is to estimate a function to predict the label from the given features. This function is called a *hypothesis*, a *model*, a *classifier* or a *predictor*. Note that a model is not only the concept for supervised learning but also for estimating distribution in unsupervised learning. This function has *parameters*, where the values of parameters are trained (learned or estimated) by using input data, and the trained hypothesis is used to predict the labels of unknown instances (or data). The first input data are called *training data* and the second one is called *test data*.

2.2.2 Estimating Parameters: Model Optimization

Estimating a model from given training data means, for example, to keep the values given by the model closest to those of training data. Thus we can set up an *objective function*, for example, in supervised learning, being consistent with the error of the model from training data, i.e. an *error function* or *loss function*. We can then minimize the objective function, which turns into an optimization problem, like minimizing the error function. The optimization in terms of training data only causes overfitting of the model to the training data. In order to avoid the overfitting, the model needs to be generalized, which is called *generalization* or *regularization*. In practice, we add one or more constraints to the objective function, where the constraints are, mathematically, terms, called *regularizers* or *regularization terms*. Thus to solve some problem, we can set up an objective function with regularizers, which we often call problem *formulation*. Usually in machine learning model formulation, model parameters to be estimated are in both the objective function and regularizers. Also model formulation sometimes has *hyperparameters*, which are given arbitrary by users or decided empirically but not trained from input data. For example, in model formulation, the coefficients of regularizers, called *regularization coefficients* are one typical hyperparameter.

Now let's get back to Example 3, where the situation is an E-commerce site. It would not be the case that an customer has bought almost all items in the E-commerce site, while usually most customers bought only just a few goods in the site. Thus the data from E-commerce, i.e. a matrix of users (rows) and items (columns), has only a small number of elements, filled by some values, showing the goods evaluation by customers or just the user purchase history. More concretely the row vectors or users have only a few number of values, with others being *missing information*. This situation of data is called *sparse* data, and interestingly sometimes this data *sparsity* is useful to solve the problem efficiently even if the given matrix is huge.

2.3 Marketing

2.3.1 Definition of Marketing

One short definition[3] of marketing is "meeting needs profitability" [56]. A more realistic way of understanding marketing is that companies, which (produce and) sell good items that people want at low price "turn a private or social need into a profitable business opportunity" [56]. The American Marketing Association gives the following longer, more formal definition: "Marketing is the activity, set of institutions, and processes for creating, communicating, delivering, and exchanging offerings that have value for customers, clients, partners, and society at large" [1].

There are two different types of definition on marketing: *social* and *managerial* definition [56]. Simply speaking, the latter is more on profitability, while the former is much less on this aspect. Readers might be likely to accept the latter

[3]We will explain several terms to be used for defining "Marketing" later.

Table 2.1: Marketing is not selling.

	Selling	Marketing
Focus	Selling	Customers
Objective	Convert products or services into cash.	Satisfy customer demands in best possible ways.
Strategy	Push: try to sell as much as possible.	Pull: capture customer demands.

definition of marketing more easily from a viewpoint of profitability. We have to point out that marketing is a broader concept than just selling/buying goods, which is only a tip of the marketing iceberg. For example, "the aim of marketing is to know and understand the customer so well that the product or service fits him and sells itself. Ideally, marketing should result in a customer who is ready to buy. All that should be needed then is to make the product or service available [29]." Simply speaking, marketing is a *customer satisfying* process, and not a process of goods producing and selling. Table 2.1 shows a summary of the difference of marketing from selling, in terms of three points: focus, objective and strategy. Regarding the focus, selling just concentrates on selling goods itself, while marketing considers more on customer satisfaction. This affects the objectives of selling and marketing. That is, the objective of selling is again selling more goods, while that of marketing is customer satisfaction, and marketing explores the best way of customer satisfaction.

In strategy, Table 2.1 shows *push* and *pull strategies* for selling and marketing, respectively. The push strategy is a promotional strategy to show the goods/products to the customers more through (for example, sales) advertisements (to take products to customers). On the other hand, the pull strategy is to pull customers in, for example, stores and web pages of the products through more customer-oriented advertisements (to pull customers to the products). The pull marketing focuses on long-term relationships with customers, by creating brand loyalty and keeping customers back, while push marketing focuses on more short-term sales. Fig. 2.7 shows schematic picture of the pull strategy and the push strategy. Again in the push strategy the retailer promotes the goods to customers straightly, while in the pull strategy, the customers are promoted to come to the retailer.

Thus *marketing management* can "view the entire business process as consisting of a tightly integrated effort to discover, create, arouse and satisfy customer needs[4]" [62]. We may then think that the social definition can be a broader idea, including managerial definition, particularly recently, and then the definition of social marketing might be the definition of marketing itself. Here a well-known definition of social marketing is as follows: "Marketing is a societal process by which individuals and groups obtain what they need and want through creating, offering, and freely exchanging products and services of value with others" [56].

[4]Maybe, instead of needs, "demands" would be more appropriate.

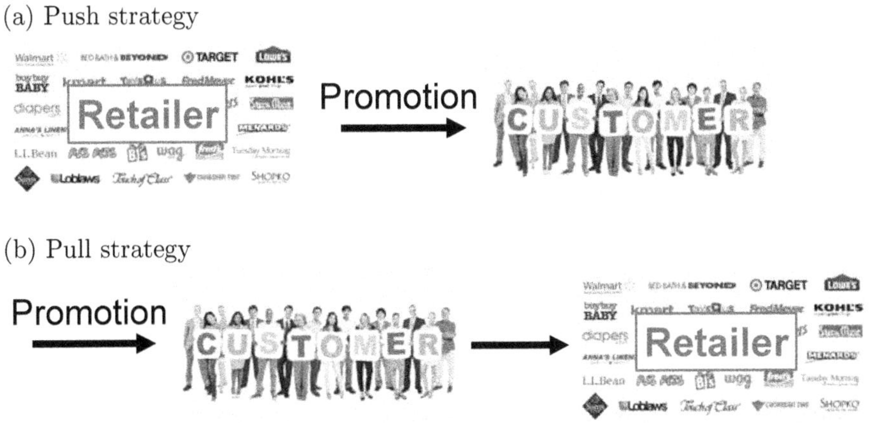

Figure 2.7: Schematic picture of (a) push marketing and (b) pull marketing.

In fact an important point of the recent marketing is that corporates treat the relationships with the society very carefully. So-called *corporate society relationships (CSR)* are now one of the major key roles of companies, and most corporates now have the devision, which is working only for CSR. That is, the companies are very cautious on the reputation, evaluation, and reviews by customers, since they directly affect the company names and brands, and eventually the sales and profits of their goods and services. In summary, now the companies have great concerns with *customer satisfaction*. This is just one aspect of societal marketing, indicating that companies and customers are connected by not only selling/buying goods and services provided by the companies but other many ways.

2.3.2 Markets

There are ten main types of entities, which can be marketed: *goods* (*items*), *services*, *events*, *experiences*, *persons*, *places*, *properties*, *organizations*, *information* and *ideas* [56]. An important point is that the marketed entities are not only goods, but also a wide variety of things which are sometimes physically seen but not necessarily so. Another point is that although there are ten types of different entities raised, we can focus on the two most abundant entities: goods and services, where goods can be seen always, while services are not necessarily.

Markets have marketers, where a *marketer* is someone who seeks a response–attention, a purchase, a vote, a donation–from another party [56]. People in markets are not only marketers, while markets have both *buyers* (or *consumers*) and *sellers* (or *suppliers*), who trade goods and/or services. Typically there are three types of markets: 1) *consumer markets*, where consumable goods and/or services are sold, mainly for individuals, 2) a counterpart is *non-profit* or *government* markets, where services are provided for customers and for example taxes are payed from customers to the non-profit organization, and 3) another counterpart is *manufacturer markets*, in which goods and/or services are traded, but buyers are not individuals but manufacturers, i.e. corporates. Also simply the term "market" is often used for specifying groups of customers in a wide variety of ways, such as foods markets (by products), kids markets (by ages or demographic data) and European markets (by locations) [56].

The physical place of the market, such as a supermarket, is called a *marketplace*, while the digital (or virtual) place of that in internet is called a *marketspace*. The *metamarket* contains both the above two concepts of markets [87, 88].

2.3.3 Marketing Concepts

Terms

We can start with terms to explain customer behaviors for marketing concepts: buyers, one of the two parties in markets, have some motivation to acquire goods or services provided by sellers. These motivation can be classified into three types: *needs*, *wants* and *demands*. Needs are the most basic requirements, such as water, foods, wears and houses, to live. On the other hand, wants are strong needs on specific goods or services. For example, cars are needs to live in the country side for transportation, while wants are on a specific car type, such as sports cars, or a specific car or car manufacturer, like Ferrari or Lamborghini. Demands are wants, where consumers with demands mean those who are able to buy the goods or services of demands, while those with wants are not necessarily able to buy the goods they want. This means that the companies might focus on consumer "demands" more than simpler consumer "wants" and also "needs". Demands of consumers for a product are measured under specified conditions, such as the consistent customer demographic feature, the specific area, etc. The measured demand is called *market demand*. This demand is real, while the estimated demand by the company side is called *company demand*. The accurate company demand (also called *future demand*) is key to success for the company.

Suppliers address to satisfy customer needs and demands. Then they develop *offerings*, which are a combination of goods and services, providing benefits, called *values*, to buyers [82]. The *customer value triad* is an idea of defining the value by a combination of three points: *quality*, *service* and *price*. Of course the value increases with quality and service but decreases with price. Marketing can be thought as creating and delivering the value to customers. *Customer satisfaction* can be a criterion to judge the value in marketing.

Marketing activities can be measured by four points: product, price, promotion and place, which are entirely called *marketing mix* or *4P* [73]. This is a rather classical measurement, and clearly focused on just selling too much. Thus several replacements have been proposed: *4C* (Consumer, Cost, Communication and Convenience) [58], where the idea is that for example, consumers should be weighed more than products, communication with customers can be more important than one-directional promotion (advertisement) and the convenience of buying products should be considered instead of place, since selling places are not important in the internet era. Another example is new 4P (People, Processes,

Programs and Performance) [56].

When we think a so-called "marketing strategy", these 4P are basis for implementing the strategy for market.

Historical Change

Now we see the marketing concepts, being divided into the following five types: production-, product-, selling- and marketing-oriented, and societal marketing concepts. The first three are rather obsolete and the last two are more modern and focused by this book.

1. **Production-oriented concept:** One straightforward idea for selling is to consider the most central people among buyers and supply products or services proper for those averaged people. Then the suppliers just seek the efficiency in time, costs and distribution. The production-oriented concept follows this idea, and also the marketing following this idea has been called *mass marketing*, typical traditional marketing. The main advantage is that this strategy can make production simple, since manufacturers or service providers can just focus on only one or a few products or services, most preferred by the major consumers. However the disadvantage is that the product or service by mass marketing may not necessarily cover a large enough part of consumers.

2. **Products-oriented concept:** The products-oriented concept is based on the idea that consumers must prefer more to the newer, better quality and higher performance. In other words, this concept is derived from only the manufacturer's view, more than the demands of buyers. Research- or technology-oriented companies sometimes make this misunderstanding.

3. **Sales-oriented concept:**
The sales-oriented concept is similar to the products-oriented concept in the sense that the concept focuses on the needs of the sellers only. However the difference is that the sales-oriented concept focuses on the sellers' needs of converting the products into cash, while the products-oriented concept focused on the product quality and performance. Thus as a result, based on this concept, companies become just aggressive in pushing their sales.

4. **Marketing-oriented concept:** The marketing-oriented concept challenges the above three concepts. In the above three concepts, companies tried to find the right customers for their products. However in the marketing-oriented concept, the companies try to produce the right products for the targeted customers. A representative marketing stranger, based on this concept, is *target marketing*, with the following three steps: 1) *Customer segmentation*: Buyers who have similar preferences in consumption or similar demographic, psychographic and/or behavioral properties can be grouped, since in reality each individual cannot be so distinguished each other and the individuals with such similar properties can be treated together in the

sense of marketing . That is, customers can be divided into a certain number of groups, which are called *market segments.* 2) *Targeting*: then sellers or marketers select the segments promising for their goods or services. This is called *targeting.* 3) Positioning: then sellers develop *offerings* for the targeted segments. This step is called *positioning.*

We will explain target marketing in more detail in Chapter 3.

5. **Societal marketing concept:** As mentioned earlier, marketing has a wider definition than just selling/buying. In particular, one important aspect of marketing is to develop long-term relationships with people in society which will eventually affect the company brands and products. The societal marketing concept indicates this type of modern marketing manner. Also a similar but rather wider concept is *holistic marketing concept*, which has four concepts of marketing: 1) *relationship marketing*, 2) *integrated marketing*, 3) *internal marketing* and 4) *performance marketing* [56].

   The most important one among the above four is 1) relationship marketing, which "aims to build mutually satisfying long-term relationships with key constituents in order to earn and retain their business [37]". A key point of customer relationship management is, being opposite to the classical mass marketing, to identify each customer and then satisfy their needs and demands. This is called *personalized marketing.* We will explain relationship marketing in more detail in Chapter 4.

   We here briefly describe the other three marketing concepts below:

   2) Integrated marketing means that different marketing activities can be assembled together, where they are strategically integrated, not just the sum of them. For example, different advertisements, such as television, radio and web sites, etc. can be integrated into a more efficient and effective way.

   3) Internal marketing means the hiring, orientation, training of company employees to keep/improve customer satisfaction. This marketing focuses on "within" companies.

   4) Performance marketing means considering not only sales profits but also wider indices, such as market share, customer loss rate and customer satisfaction and other measures, which are eventually the evaluation of the company in terms of social responsibility.

Again this book focuses on two marketing strategies: target marketing and relationship marketing, which are derived from the above **4. Marketing-oriented concept** and **5. Societal marketing concept**, respectively. Also we describe target marketing and relationship marketing in detail in Chapters 3 and 4, respectively.

The most important point for the company growth is to discover and foster the competences to boost and protect the company business. These competences are called *core competencies*, which should have the following characteristics [79]: core competencies 1) highly contribute to achieve the customer benefits, 2) can be applied to not only one but many markets and 3) are hard to be imitated.

Chapter 3

Target Marketing

Target marketing is already an established approach, which has now been considered as the most standard marketing strategy. This strategy focuses more on starting selling new goods (products) or attracting new customers, comparing with relationship marketing, which focuses on retaining the existing customers for the existing products. We will explain relationship marketing in the next chapter, i.e. Chapter 4.

Target marketing has three steps: market 1) Segmentation, 2) Targeting and 3) Positioning, which is totally called the *STP strategy* [30, 112]. Fig. 3.1 shows a schematic picture of the STP strategy: (a) Segmentation: customers are segmented into a certain number of groups, in which customers share the same wants, needs and preferences, (b) Targeting: out of the segmented groups, business-wise most promising targets for the company are selected and (c) Positioning: the company considers possible offerings for each of these target segments.

We will explain each of these three steps below.

3.1 Market Segmentation

In this step, a market is divided into distinct groups of buyers, where buyers in each group have common needs, characteristics, or behavior, resulting in that they must prefer similar products, and also these buyers are distinct from buyers in different groups in terms of their wants, needs and marketing mixes [5]. In reality we can use the following information for segmenting the market, which can be two types: customer market and business market. First, for customer market, the following four viewpoints are raised as bases (features) to perform market segmentation [56]:

1. **Geographic segmentation (where?)**
   Information on the place the customer is living: region (continent, country, state or even neighborhood), size of area, population size, population density (urban, suburban or rural), climate, etc.

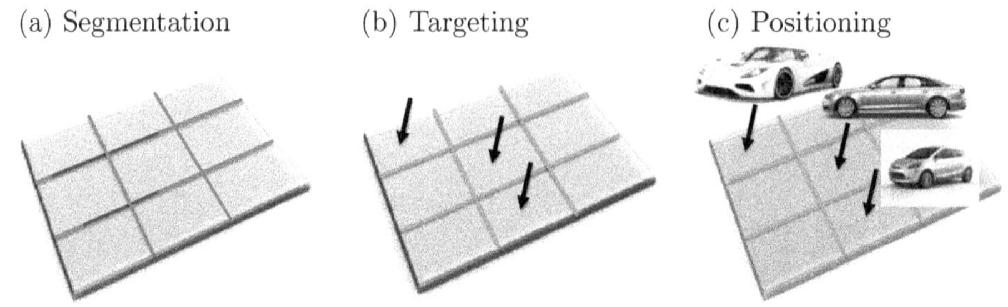

Figure 3.1: Schematic pictures of the STP strategy: (a) segmentation (b) targeting and (c) positioning.

2. **Demographic segmentation (what?)**
   Basic variable of customers: age, gender, life stage (like staying a single or being married), family size, family life cycle, generation, income, occupation, education, ethnicity, nationality, religion, social class, etc.

3. **Psychographic segmentation (who?)**
   Psychological traits of customers: personality, lifestyle (activities, interests, opinions, attitudes) and values.

4. **Behavioral segmentation (how?)**
   Customer behavior towards the product: knowledge of the similar products, usage rate of the products, user status (potential, first-time, regular, loyal), readiness to buy, occasions (holidays and events that stimulate purchases), etc.

Note that entirely they are customer (user) profiles, and currently they are collected through internet, such as customers' online activities [104], and this collection and generation of user profiles is called *user profiling*.

For business (manufacturer or corporate) market, we can use the same four segmentations for customer market. Additionally considering a rather special situation of business market, the following four viewpoints can be used as bases (Note that for business market, a customer means a company (business partner)) [56].

5. **Operating variables:** Customer technology, user or nonuser status (heavy, medium, light or non users) and customer capabilities.

6. **Purchasing approaches:** General purchasing policies, purchasing criteria, nature of existing relationship, etc.

7. **Situational factors:** Urgency, specific application and size of order.

8. **Personal characteristics:** Buyer-seller similarity, attitude toward risk and loyalty.

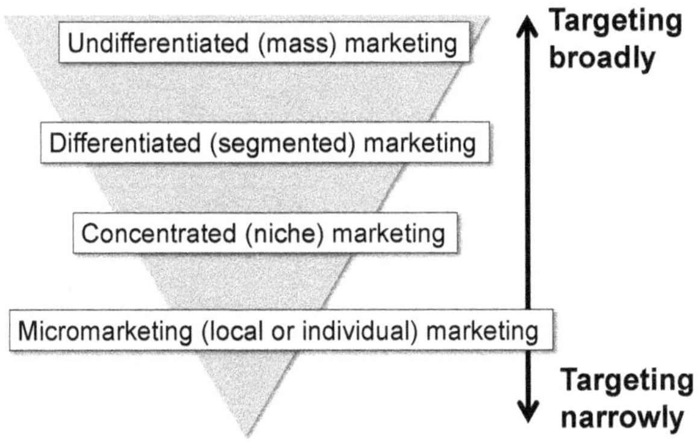

Figure 3.2: Various scales of targeting: from mass marketing to micro marketing.

Then using these information, a market should be divided into a certain number of segments, considering the following criteria:

1) **measurable:** Each segment should be quantitatively measured, by which one segment can be compared with another segment.

2) **substantial:** Each segment should be large enough, while the buyers in each segment should be consistent enough. By doing this, one segment can be clearly distinguished from another segment.

3) **accessible:** Each segment can be sufficiently reached.

4) **differentiable:** Each segment should be internally homogeneous and externally heterogeneous, such that one segment is comprehensible and explainable by some reason.

5) **actionable:** Each segment should be set so that possible business offering can be built for the segment.

Over all the point on the manner for segmenting customers is that each segment should be reasonably distinguishable from another segment always.

3.2 Market Targeting

Targeting has two steps: 1) segment evaluation and 2) selection of segments.

3.2.1 Segment Evaluation

We can evaluate the differences among market segments. In fact this step might be a key in the STP strategy. In order to evaluate the segments effectively, we can focus on the following three factors of each segment:

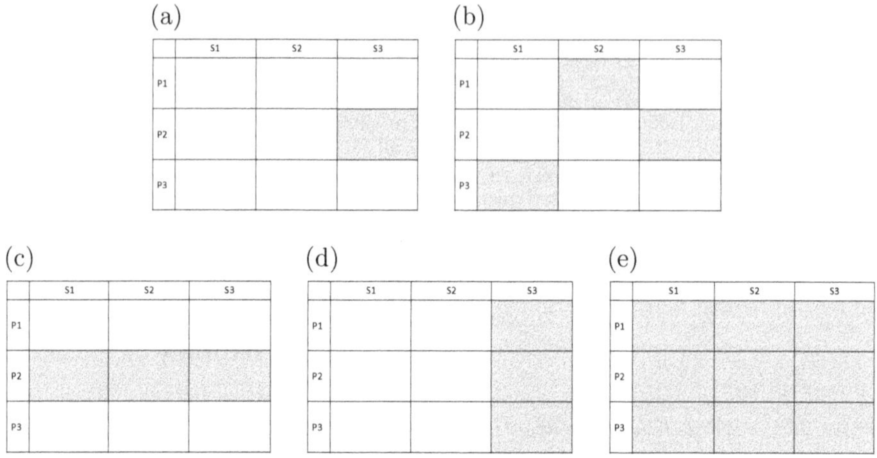

Figure 3.3: Five types of segment selection: (a) single-segment concentration, (b) multiple segment specialization, (c) product specification, (d) market specification and (e) full market coverage.

1. **Segment features** We can evaluate the objective features of each segment, such as sales, profitability and growth rates, etc.

2. **Segment attractiveness** We can further evaluate the attractiveness of each segment, such as level of competition, substitute products, buyer purchasing power and impact of competitors, etc.

3. **Company features** The company itself has its objectives and resources, and it should be considered and measured if each segment matches them well.

3.2.2 Segment Selection

We can decide which and how many segments we should choose for business. The above step of segment evaluation is important for choosing the segments correctly. When segments are selected, we can consider the scale of segments which the company will choose. Fig. 3.2 shows the difference of the targeting scale, which has mainly four different size (from larger to smaller) in the scale: undifferentiated (mass) marketing, differentiated (segmented) marketing, concentrated (niche) marketing and micromarketing (local or individual).

Also when we select the segments, in general there are five ways of choosing the segments: 1) *single-segment concentration*, 2) *multiple segment specialization*, 3) *product specialization*, 4) *market specialization* and 5) *full market coverage*. Fig. 3.3 shows all five types of selecting segments, where we assume that market (customers) are grouped into segments by only two viewpoints: products (three categories: P1, P2 and P3) and market (three categories: S1, S2 and S3), resulting in a matrix of nine (= 3×3) segments. As shown in Fig. 3.3, simply (a) single-segment concentration focuses on only one segment, while (b) multiple segment specialization takes more than one segments. First, single-segment concentration means that the company goes into a *niche* market (see Fig. 3.2). As the segment is smaller, the ultimate segment has only one individual, which is called *customized marketing* or *one-to-one marketing*. On the other hand, in general companies are likely to take multiple segments rather than only one segment, considering the synergies among multiple segments, which share some similarity. The taken multiple segments are called a *supersegment*. Also when taking multiple segments, the company might want focusing on only one type of product or one type of subcategory of the market, which are (c) product specification and (d) market specification, respectively. In fact these two can be thought as special cases of multiple segment specialization. Finally we can take all segments, resulting in (e) full market coverage, which corresponds to mass marketing.

3.3 Market Positioning

Positioning means developing proper offerings so that each new offering can occupy the corresponding, selected segment. In fact, positioning cannot be independent of the market targeting in Section 3.2, since the selection criterion of segments would be based on what offerings can be provided by the company. We here just consider market positioning, rather being independent of the segment selection, like that we need develop some new offering in an already given segment.

Market positioning, i.e. building a strong offering in a selected segment, is a very similar process to establishing a *brand* of the product or the company. Thus this selection would be useful to understand the process of generating a proper brand.

For developing a competitive offering, clearly a key point is to understand the environment of the product, such as specifying the players/products (or already existing offerings) in the segments, which might be selected. This process will make design some competitive offering of the product definitely easier. Thus we start with explaining the way to analyze and understand the environment of the product.

3.3.1 Identifying Competitors: SWOT Analysis

A well-established approach for understanding the environment of a product or a company is so-called *SWOT* (Strengths, Weaknesses, Opportunities and Threats) analysis [59]. The SWOT analysis would be the most well-used market environment investigation method under a lot of aspects of marketing and management when examining the environment in which the product or the company is placed. Thus this analysis is always written in any book on marketing.

Table 3.1 shows the most abstract view of the SWOT analysis, which can be depicted as a 2×2 contingency table. The SWOT analysis first divides the environment (rows) into two cases: 1) the first is the internal environment,

Table 3.1: SWOT Analysis. The SWOT analysis shows the Strengths and Weaknesses (of the product or company) under the internal (controllable) environment, and Opportunities and Threats under the external (uncontrollable) environments. Table 3.2 shows general and more real contents of each of Strengths, Weaknesses, Opportunities and Threats.

	Good effect	Bad effect
Internal (Controllable)	Strengths	Weaknesses
External (Uncontrollable)	Opportunities	Threats

Table 3.2: A more real example of SWOT Analysis. That is, this table shows general and more real contents of each of the Strengths, Weaknesses, Opportunities and Threats.

Strengths	**Weaknesses**
Intellectual property (IP)	Expiring IP
Cost advantages	Rising costs
Skilled workforce	Unskilled workforce

Opportunities	**Threats**
New technology	Emergence of new competitors
Relaxing government regulations	Pending government regulations
Elimination of international trade barriers	Increased trade barriers

which is controllable, and 2) the second is the external environment, which is uncontrollable. Then for each of these two types of environments, we can think about the pros (good) and cons (bad) of the product or company which become *strengths* and *weaknesses*, respectively, under the internal environment and also *opportunities* and *threats*, respectively, under the external environment.

Table 3.2 shows example remarks for each of the four aspects of the SWOT analysis for the environment of the selected segment. In fact this observation by SWOT analysis gives a clearer picture of the advantage and disadvantage of the product, service or company. Also at the same time, these results would assist developing new offerings easier.

3.3.2 Understanding Competitors: Perceptual Map

By the SWOT analysis, we can raise the features of the product (or the company), which are divided in the four aspects, as shown in Tables 3.1 and 3.2. These are features obtained by comparing a particular product with others (competitors) in

Table 3.3: Required data for SWOT Analysis.

	Internal (Controllable)	External (Uncontrollable)
Good effect	Strengths	Opportunities
Bad effect	Weaknesses	Threats
	Customer	Secondary data
	Employee	Environment
	Capital	Industry
	Resources	Competitor
	Process	
		Customer feedback

Table 3.4: Feature examples for drawing the perceptual map.

Sample category	Feature example
Digital camera	memory size, picture quality, robustness, size, weight
Instant coffee	country of origin, flavor, quality, taste
Soft drinks	energy boost, for kids/females/males, healthy, high/low in caffeine, high/low in sugar, old-/new-fashioned

the same sector (segment), while these features are obtained rather subjectively in the SWOT analysis, in the sense that they are not necessarily derived from data. Thus instead of using the idea of the SWOT analysis, it would be better to think about a more objective manner to understand the environments under which the product is placed. A *perceptual map* [41] would be useful for realizing this purpose, making the comparison between the product and its competitors clearer.

The perceptual map (or *positioning map*) is "visual representations of consumer perceptions and preferences. They provide quantitative portrayals of market situations and the way consumers view different products, services, and brands along various dimensions" [56]. Practically a perceptual map has a space with a few number (typically two) of axes, i.e. a low-dimensional (again two-dimensional) space, in which products are distributed, showing the similarity/difference between products by the distance between products. Seeing examples would be good to understand the perceptual map intuitively, and readers may take a look at Fig. 3.4 briefly first.

The perceptual map is generated from a matrix of (products (competitors) $\times$ features). Table 3.4 shows practical, sample features (second column), which can be raised for sample products and their segments (first column). They are just samples, and depending on products and segments, these features can be or must be changed. For example, in this table, if the focused product and its competitors are digital cameras (i.e. the digital camera section), possible features will be the memory size, picture quality, robustness, size, weight, etc. These examples are

Table 3.5: (top) Original survey data for generating a perceptual map. (middle) An matrix obtained by switching the first and second columns of the top, and the rows are reordered by the first column (products) and then the second column (customers). (bottom) (Product × feature)-matrix, which is transformed from the middle matrix, by summarizing consumer responses for each product (competitor) into one vector.

		Feature 1	Feature 2	Feature 3	...
Customer1	Product A				
	Product B				
	...				
Customer2	Product A				
	Product B				
	...				
Customer3	Product A				
	Product B				
	...				

⇓

		Feature 1	Feature 2	Feature 3	...
Product A	Customer1				
	Customer2				
	...				
Product B	Customer1				
	Customer2				
	...				

⇓

	Feature 1	Feature 2	Feature 3	...
Product A				
Product B				
...				

rather a list of specs of digital cameras, while features can be obtained from user (customer) evaluation, i.e. survey (or questionnaire) data. In particular, rather than just catalog specs, survey from customers would be practically more useful and comprehensible.

Suppose that the matrix of products × features for a perceptual map is generated from customer survey data, we note that this original data has multiple customers, where each customer provides evaluation on multiple products. The top table in Table 3.5 shows a schematic example of this original survey data, which has multiple products (row vectors) by each customer. Then to generate a perceptual map, we need transform this original survey data into a matrix of (products (competitors) × features). For this purpose, we need summarize multiple vectors for each product by multiple customers into one vector, as shown in the bottom of Table 3.5. More in detail, we, for example, take an average of feature values, for one feature, of all vectors (multiple customers), for the same product over all customers, resulting in one vector for one product. The bottom of Table 3.5 shows a schematic example of a matrix of (products × features).

The perceptual map can be divided into two formats: 1) two features and 2) multiple features. That is, the original features, which are schematically shown in Table 3.5, might be reduced into a smaller number of features, or some of the original features might be selected. This corresponds to Step 1 below. In both cases of two and multiple features, the main procedure of generating the perceptual map is common and given as follows:

Step 1. Selecting features. If we use only a limited number of features to generate a perceptual map, we need to select features out of all given features, while if we use all features to show a perceptual map, we can skip this step. The features we can select would be those most useful to understand the relationships between the product and the competitors. In other words, we should select features most closely connected to the product purchase decision of consumers. Again the selected features should be used by consumers to differentiate possible given offerings.

The number of features to be selected is decided by the format of the perceptual map to be drawn. That is, for example, the number becomes two for the perceptual map with two features.

Another point, which is not mentioned so much in the literature of marketing, is that the features to be selected should not be so relevant each other but more irrelevant or independent of each other. Mathematically they should be orthogonal each other to differentiate the competitors to be plotted on the perceptual map most efficiently.

Step 2. Selecting competitors against the product.
The next step is to list up all competitors against the product (or brand (or company)) which are compared with each other in the same category. Then all competitors with the product are plotted over the space generated by the features selected in Step 1.

Step 3. Quantify the competitors. We need quantify each competitor to plot it on the space by the selected features. The most primitive way is to give each competitor a score manually. Another possible way is to generate the scores by using survey data, i.e. the customer answers of questions on the selected features. Simply the survey asks consumers about their scores on each of the selected features. For example, regarding digital camera, each consumer will answer their preference or understanding of the camera picture quality of the product and also each of the competitors. Again the top of Table 3.5 shows such original survey data, and we need to generate the data for a perceptual map from the top of Table 3.5. For example the simple mean over the answers given by customers can be computed as a score for the corresponding feature. The bottom of Table 3.5 shows a schematic example of a matrix of (products × features).

Table 3.6: (top) A (customers × features)-matrix, which is from the original survey data (the top of Table 3.5). (middle) Customers can be grouped into several segments, by market segmentation already (Note that the segments were generated by different features from the features here) in Section 3.1, meaning that customers can be grouped and reordered, according to segment. (bottom) We can compute CSIP (customer segment ideal point) of each segment of customers, resulting in a (CSIPs × features)-matrix.

	Feature 1	Feature 2	Feature 3	...
Customer1				
Customer2				
...				

⇓

		Feature 1	Feature 2	Feature 3	...
Segment i	Customer3				
	Customer5				
	...				
Segment ii	Customer1				
	Customer7				
	...				
Segment iii	Customer2				
	Customer6				
	...				

⇓

		Feature 1	Feature 2	Feature 3	...
Segment i	CSIP i				
Segment ii	CSIP ii				
Segment iii	CSIP iii				
...					

Step 4. Plot the scores over the feature space. Finally the computed scores of the selected features are plotted on the same feature space on a perceptual map. Note that each competitor becomes one point on the space.

Thus the perceptual map has points of products (or companies) or competitors. We here note that at the same time, we can generate a matrix of (customers × features) from the original survey data. The top of Table 3.6 shows a schematic picture of a (customers × features)-matrix, which can be generated from the original survey data, say the top of Table 3.5, a schematic table of original survey data, by taking the average/mean of all products for each customer. Then consumers who answered the questions in the survey data can be divided into groups, according to the segments already generated in market segmentation, which was explained in Section 3.1. Furthermore the mean/average over the scores by the consumers in each segment can be also a point for each segment on

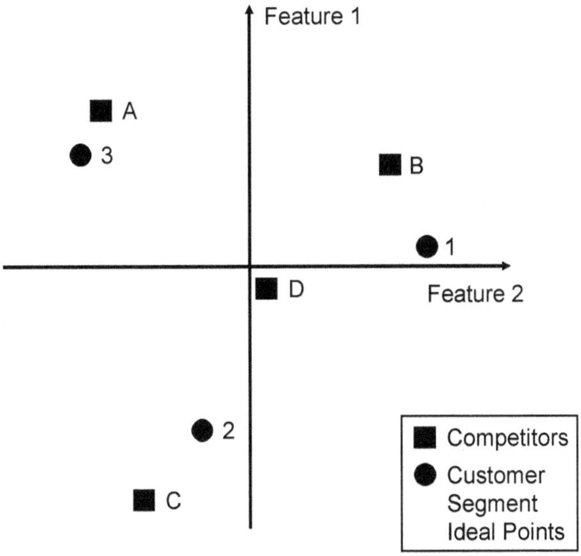

Figure 3.4: Example of a perceptual map with two features.

the feature space. We call this point the *customer segment ideal point* (CSIP) for each segment. Of course, one possible problem setting is to estimate CSIP directly from data, such as survey data [25]. We can plot these points of segments over the feature space. Thus, a perceptual map can have two different types of plots: competitor (products/brands/companies) points and the customer segment ideal points[1].

Perceptual Map with Two Features

Fig. 3.4 shows a schematic example of a perceptual map with only two features which results in a two-dimensional (2D) space. As mentioned above, a perceptual map has two types of points: 1) competitors and 2) customer segment ideal points, which are shown by ■ and ●, respectively, in the 2D space.

After drawing a perceptual map, we can practically have several findings/interpretation from a perceptual map. Then we show several examples, which can be derived from Fig. 3.4.

1. **Distinct positioning of competitors.** Fig. 3.4 shows all four competitors are well-distributed and positioned away each other. Thus in Fig. 3.4 we can say that there are no highly competitive situations, in which two or more competitors are plotted together in a very small region (That is, if two

[1] In the literature, these two types of plots are sometimes separated. For example, perceptual maps with both types are called *joint maps*, while perceptual maps with competitors only are still called perceptual maps.

or more competitors are positioned very close each other, we can see their positioning is highly competitive each other).

In Fig. 3.4, to understand the competitive situation better, we can replace either of the two features with another feature, so that some competitors can be located very closely.

2. **Match of competitors against segments.** In Fig. 3.4, product A is very close to segment 3, expecting that product A should be well accepted by the customers of segment 3.

   Also products B and C are rather close to segments 1 and 2, respectively, indicating that these products are not necessarily perfect but rather consistent with customer preferences of segments 1 and 2, respectively.

   On the other hand, product D is rather far away from any segment and also no strong trends regarding both two features 1 and 2, implying that no market segment corresponds to product D. In other words, positioning of product D is not necessarily well made.

3. **Market gap or incomplete map?** In Fig. 3.4, the right-bottom side is rather empty. This type of space can be caused by either/both of the two reasons:

   Market gap: In the market, there are no products or companies which satisfy the market needs, corresponding to the empty part. In particular if there is an empty area which does not have any product but a CSIP, this indicates a good chance to develop a product which meets the preferences/demands of the customers who are represented by the corresponding CSIP.

   Incomplete map: Maybe the empty space just means the lack of products which should be examined but not listed up when the map is generated.

4. **Product development/improvement strategy.** Again in Fig. 3.4, the position of product D was away from any CSIP, i.e. any of CSIPs 1 to 3. This indicates that product D has no good values in terms of customer preferences in the both features, i.e. features 1 and 2. In order to improve the virtue of product D, there are two possible ways in Fig. 3.4:

   D → 1: The value of feature 2 of product D can be changed as closer to that of CSIP 1. This means that the value of feature 2 can be made close to that of CSIP 1, keeping the same as the value of feature 1.

   D → 2: Similarly, the value of feature 1 can be made closer to that of CSIP 2.

Multiple (More Than Two) Features

We can increase the number of features selected from given data. In this case there are two ways to represent multiple features.

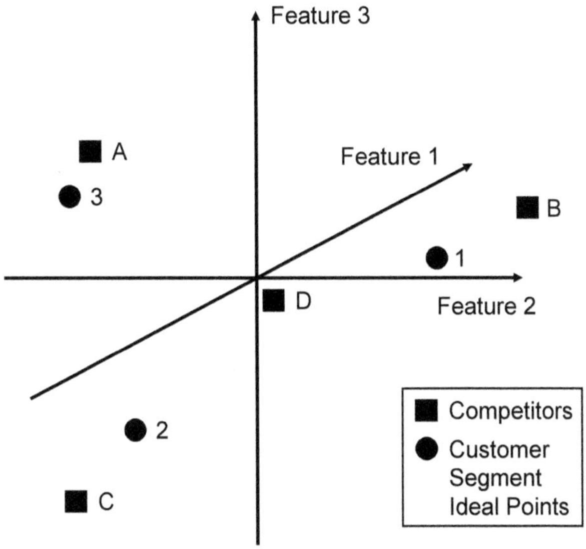

Figure 3.5: Example of a perceptual map with three features.

1. **Simple increase of dimensions:** One way is to increase the dimension of the space, as the number of the selected features increases. Fig. 3.5 shows an example of a perceptual map with three features. If the number of features is still three, the three-dimensional perceptual map in Fig. 3.5 is visually understandable. However if the number of features is larger than three, this high-dimensional perceptual map cannot be visually understandable. Thus this simple approach is unavailable.

2. **Mapping multiple features over the 2D space**
   Another visualization with more than two features is to show all on the 2D space. That is, competitors, customer segment ideal points and (high and low values of) features are presented on the 2D space. Fig. 3.6 shows a schematic example of this visualization. You can see that Fig. 3.6 has multiple features with their (high and low) values, corresponding to several different parts of the 2D map. Briefly the idea behind this mapping is to generate two new features from all original features. Then when we generate the new features, we can have two different ideas: 1) on the 2D space of the two new features, competitors can be distributed as diversely as possible. 2) similarly on the 2D space of the two new features, competitors should be as closely positioned as possible.

   For each of the above two purposes, we can raise a multivariate analysis approach. In fact this is based on machine learning already, and so more details of this approach will be described in Section 6.4.2.

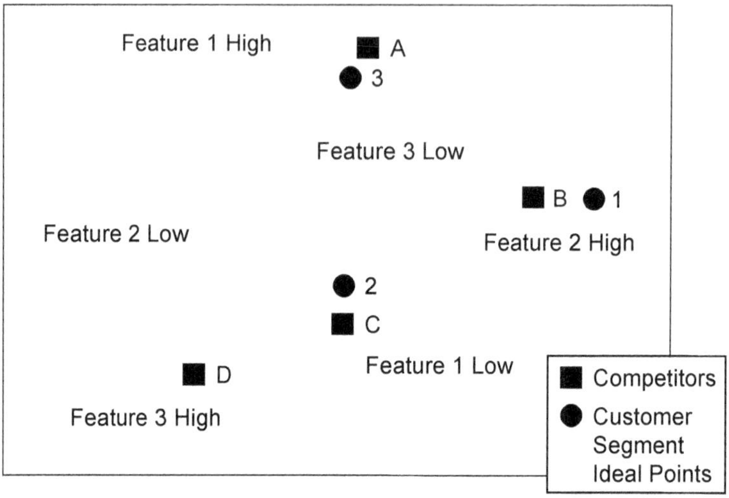

Figure 3.6: Example of a perceptual map with three features on a 2D space.

Further Use of Perceptual Map in Marketing

So far perceptual mapping has been used for placing products (or services) as well as customers (note that customers are represented by ideal points of customer segments) on (mainly) a 2D space generated by their features. This means that perceptual maps are for visualizing products and customers. Visualization by perceptual maps is useful and can be used in other domains of marketing. One example can be found in the process of developing a new product. That is, when new products (or services) are developed, we need to generate new-product ideas, which can be then examined through product *concepts*. We note that consumers do not buy ideas but products, based on the concept for the products or the concepts themselves. Thus after generating a new idea, we first generate the perceptual map (product positioning map) of the products related with the new idea to check the positioning of the new idea against the existing, competing products. We then switch the perceptual map on products to that of *brand concepts* which is called a *brand positioning map*. Thus new ideas (or concepts) can be compared with and positioned against other existing brand concepts in a brand positioning map so that the new concepts can be evaluated in terms of the positioning against existing product-brands.

In fact brands as well as products and consumers are just examples, and perceptual maps can be used for understanding any environment under which the targeted items are placed in marketing always.

Chapter 4

Relationship Marketing

Relationship marketing is centered on customers more[1]. *Customer Relationship Management* (CRM) is a time-series management process of attracting new customers and keeping existing customers to maximize their loyalty to the product. In particular, CRM is more focused on retaining the existing or loyal customers of the product, more than finding and inviting new customers to the product. We start explaining the idea and objective of relationship marketing, being followed by *retention marketing*, which is the essential part of relationship marketing or sometimes the relationship marketing itself.

4.1 Customer Relationship Management: Idea and Objective

Table 4.1 shows several typical features of (customer) relationship marketing, comparing against the traditional mass marketing. From this table, we can see that not only the point of focusing on individual customer simply but also the following at least three key points of relationship marketing exist:

1. **Profitable customer:** As shown in Table 4.1, one clear contrast between relationship marketing and mass marketing is profitability. That is, CRM focuses on customers, i.e. *profitable customers* (see later), while the focus of mass marketing is on products or brands.

2. **Loyalty to the product:** The objective of the relationship marketing is to maximize the *loyalty* of each profitable customer to the product. Thus, customer relationships are also called *loyalty relationships*.

3. **Long-term value:** Also this loyalty must be maximized from a *long-term* viewpoint, simply life-long time span of the customer. This is not like maximizing the amount/ profit of one time purchase, i.e. some short-time span of profit.

[1]The view change on the relations with customers is well described in [61]

Table 4.1: Relationship marketing vs. mass marketing.

	Relationship marketing	Mass marketing
Communication	Interactive communication with each individual.	Use mass media to build brand and announce products.
Idea	Collaborate with customers.	Sell to customers.
Goal	Customer and employee satisfaction.	Company satisfaction.
Profitability	Profitable customer.	Profitable products and brand.
Relationship	One-to-one.	One-to-many.
Share	Share of customer (wallet share).	Market share.
Target segments	Own-label product range.	Middle-class.
Marketing	Customer retention.	Customer acquisition.
Time range	Long-term customer engagement.	Short-term goal.
Value time	Lifetime value.	One time value.

As shown in the table, these three are totally opposite ideas of mass marketing, which focuses on average people in the short-sited view. This is because the objective of mass marketing is to minimize the product manufacturing cost of the company.

We then explain more on these three points: 1) profitable customer, 2) customer loyalty (satisfaction) and 3) long-term value, in the subsequent two subsections.

4.1.1 Profitable Customer

Profitable customers can be defined as those who provide constant revenue to the company, more than the cost that the company has to spend for the customers. Profitable customers can be detected by profitability of each customer by following the above definition. Detecting profitable customers leads to customer profitability analysis, which clarify the products purchased by customers.

Fig. 4.1 shows a sample, schematic picture of a table on customer-product relations, by using four types of products and three types of customers, where each entry is the number of products purchased by the corresponding product and customer. In this table, customers are classified into three types (high-profit customers, mixed-profit customers and losing customers) by their profitability computable by using the definition and also the purchase data of the customers. Also products can be classified into four types (highly profitable products, profitable products, unprofitable products and highly unprofitable products) by profitability. Then in each column we can see the number of entries for each row, by

	C_1	C_2	C_3	
P_1				Highly profitable products
P_2				Profitable products
P_3				Unprofitable products
P_4				Highly unprofitable products
	High-profit customers	Mixed-profit customers	Losing customers	

Figure 4.1: Schematic table of four types of products × three types of customers (from [56]).

which we can see what products are purchased by what types of customers. In fact highly profitable products would be purchased by high-profit customers, and vice versa. Thus we can see efficient relations and inefficient relations, where the former is highly profitable products purchased by high-profit customers, while the highly unprofitable products purchased by only losing customers.

Then we can consider what we can do for the customers in C_2 and C_3, which are mixed-profit and losing customers, respectively:

1. We might raise the price of the products these customers are purchasing.
2. We might remove the products, having been purchased by these customers.
3. We can recommend and sell customers in C_2 and C_3 highly profitable products.

4.1.2 Extracting Profitable Customers – RFM Analysis

We intuitively explain profitable customers by using the table in Fig. 4.1, while we have not necessarily defined the profitable customers clearly. However we can define and detect profitable customers more clearly by using the given historical purchase data of a number of customers. This approach is close to data-driven automatic detection of profitable customers, i.e. a machine learning-based technique, while we are still unable to say this is a machine learning approach. We now raise one technique, called *RFM* analysis, for extracting profitable (or loyal) customers. This approach evaluates customers from the following three different viewpoints on purchasing goods:

R – recency: The *recency* indicates the duration between the current date and the last date, on which the customer purchased some item.

F – frequency: The *frequency* indicates how many times the customer has purchased the good within a certain amount of period, say one year.

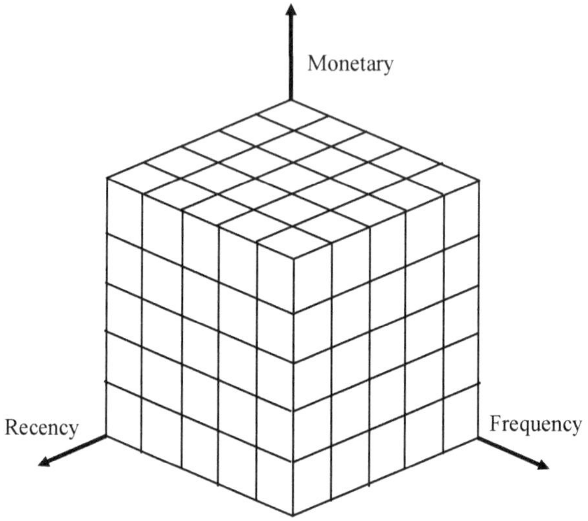

Figure 4.2: RFM analysis: segments customers in terms of three viewpoints: recency, frequency and monetary.

M – monetary: The *monetary* indicates how much amount the customer has purchased the good within a certain amount of period, also say one year.

Then each of the three features is first segmented into a small number of classes, say five, according to the amount of each viewpoint by all customers. The simplest way is, for each feature, to generate several equal-sized bins, say five bins, which can be five segments and then scored, like 1 to 5. A more sophisticated way is to assign some distribution which can be estimated from the real amount by customers. Then again this distribution is segmented into several groups, say five groups, according to, for example, the number of customers. Finally by using all three features, all customers can be segmented into 125 (=5 × 5 × 5) groups.

After finishing this segmentation, each customer can be assigned to one of the segments made, such as (3, 2, 5). For example, for some customer, let s_R, s_F and s_M be the score (or the assigned class) of the three viewpoints, i.e. recency, frequency and monetary, respectively. Then the entire score for this customer can be computed by using the linearly weighted score over the three features, as follows:

$$s_{\text{all}} = w_R s_R + w_F s_R + w_M s_M \quad (4.1)$$

$$= \sum_{k \in \{R,F,M\}} w_k s_k, \quad (4.2)$$

where w_k is the weight for each viewpoint k. This weight w_k would be able to be estimated from data but can be generated by some expert, thinking about what viewpoint is more important for extracting profitable customers for the company than others. However if no such prior knowledge, the weights can be just uniform, like all one, meaning that scores from the three features can be just summed up. In fact currently in literature of marketing, a standard manner is just using uniform weights.

Fig. 4.2 is a schematic picture of RFM analysis, where customers are segmented into five groups by each of the three viewpoints, i.e. recency, frequency and monetary, resulting in 125 (=5 × 5 × 5) groups, from which we can select profitable customers, who can achieve high (highest) scores by (4.2). The group located most far from the origin would be one of the most paid-attention group as profitable customers.

At the same time, by using RFM analysis, we can focus on some particular group out of all segmented customers (say 125 groups in Fig. 4.2), and so we can apply this analysis to customer segmentation as well. In more detail, there are some segments already empirically pointed out, in the result by the RFM analysis, for example, as follows:

Champions: They are simply the customers with the highest scores in (4.2).

Loyal customers: They give high scores in monetary, regardless of the values of recency and frequency.

Potential loyal customers: They give middle to high scores in monetary, while they are with high recency and also high frequency.

We note these labels of customers are connected to "retention marketing" (Section 4.2) and "marketing funnel" (Section 4.2.1), in the sense that customers in each stage of the marketing funnel can be labeled by RFM analysis or some more general and similar approach to RFM analysis. An intuitive example of the customer label is the "loyal customers" and "potential loyal customers", etc. in the above list. Again as a conclusion, RFM analysis is a standard method for capturing profitable customers and grasping the nature of customers in given data.

4.1.3 Customer Loyalty Satisfaction

Building tight loyalty relationships to customers is the essence of all business. In order to create the loyalty relationships, *customer satisfaction* is the point to be paid attention most. In general highly satisfied customers would stay as loyal customers, by buying products introduced by the company more and paying less attention to price and other competing brands. Thus the company needs constantly provide satisfaction to customers, by keeping the product performance more than the expectation of the customers. On the other hand, one small item of note is that customer satisfaction does not necessarily mean customer loyalty always. In particular low level satisfaction would not always connect to loyalty, and middle level satisfaction might not be good enough to make the customers stay buying the same product repeatedly for a long time. However highly satisfied customers would be likely to repurchase the goods they prefer and evaluate the goods in a highly positive way.

Also customer satisfaction is important not only from loyalty but also the viewpoint of societal marketing, since currently, as briefly mentioned in Chapter 2, companies need care the product reputation, review, etc, which can be easily spread through a lot of ways, such as social networking services (SNS), in the modern era of internet.

4.1.4 Customer Lifetime Value (CLV)

Again relationship marketing considers not short-sided, say one time, selling, but long-term, lifelong time span. That is, the company needs the *profitability* of customers not by only a small number of times but a certain long time span, especially *lifetime*.

We now show the way to compute lifetime customer profitability, which is called *customer lifetime value* (CLV). There are some models for CLM, such as that considering the shareholder values to be affected by CLV [11]. We however show rather the simplest model to consider CLV [20, 38].

Let F_t be the profit (called *margin*)[2] of the product for time t and r_t $(0 \leq r_t \leq 1)$ be the probability (called *retention rate*: see the footnote in detail)[3] that the customer is back to buy the same product again at time t. Also $1 - r_t$ is the probability called *churn rate*.

We place two assumptions over the margin of the product and also the retention rate: 1) F_t is constant over t, i.e. $F = F_1 = F_2 \cdots = F_T$, and 2) the retention rate is also consistent over t, i.e. $r = r_1 = r_2 \cdots = r_T$. Then, the customer value at time 0 can be $F_0 = F$ and that at time t follows a geometric sequence over t, as follows:

$$\text{Customer value}(t) = Fr^t \tag{4.3}$$

In order to obtain CLV, we can first sum the customer value over t from 1 to T, as follows:

$$\sum_{t=1}^{T} \text{Customer value}(t) = \sum_{t=1}^{T} Fr^t = Fr \sum_{t=1}^{T} r^{t-1} = Fr\frac{1-r^T}{1-r}. \tag{4.4}$$

CLV is the case of $T \to \infty$ in (4.4) and since $0 \leq r \leq 1$,

$$\text{CLV} = \frac{Fr}{1-r}. \tag{4.5}$$

r is the retention rate and $1 - r$ is the churn rate. Thus (4.5) shows that CLV is obtained by multiplying F by the odds between the retention rate and the churn rate.

[2]In machine learning, the term "margin" generally indicates the distance between two different classes in supervised learning (Section 5.2.1), while the profit at some time point is called the *margin* in one aspect of considering customer lifetime value

[3]CLV can be applied to collective groups to which each customer joins. Typical examples of such groups are mobile phone customers, airline frequent flyer program members, SNS users, etc. In this case, the *retention rate* means the probability that the customer stays in the group as a member, and $1 - r_t$ is the probability that the members (customer) left from the group. This probability is called churn rate.

This simple model, (4.5), can be extended by a variety of minor modifications. For example, we can assume that the original F_t can be discounted over t as follows:

$$F_t = F \cdot d_t, \tag{4.6}$$

where d_t is a discounted coefficient, satisfying $0 \leq d_t \leq 1$. That is, F_t should be decreasing as time t goes by. One example we can take for d_t is as follows [12]:

$$d_t = (\frac{1}{1+d})^t. \tag{4.7}$$

Then from (4.4), (4.6) and (4.7),

$$\sum_{t=1}^{T} \text{Customer value}(t) = \sum_{t=1}^{T} F(\frac{1}{1+d})^t r^t = F\sum_{t=1}^{T}(\frac{r}{1+d})^t \tag{4.8}$$

$$= F(\frac{r}{1+d})\sum_{t=1}^{T}(\frac{r}{1+d})^{t-1} \tag{4.9}$$

$$= F(\frac{r}{1+d})\frac{1-(\frac{r}{1+d})^T}{1-\frac{r}{1+d}}. \tag{4.10}$$

We can set $T \to \infty$ to obtain CLV and by using $0 \leq \frac{r}{1+d} \leq 1$,

$$\text{CLV} = F(\frac{r}{1+d})\frac{1}{1-\frac{r}{1+d}} \tag{4.11}$$

$$= \frac{Fr}{1+d-r}. \tag{4.12}$$

Still this is simple, while this is the most well used metric for CLV [33]. Also another practical modification is that we could subtract the original production cost of the product from (4.12), as follows:

$$\text{CLV} = \frac{F \cdot r}{1+d-r} - \text{acquisition cost}. \tag{4.13}$$

The F and d can be computed for the product, regardless of customers, and so the retention rate r can be only the value, which is obtained from customers. Looking closely at (4.12), if the retention rate is higher (closer to 1), the numerator is larger and the denominator is smaller, resulting in CLV is higher. Thus assuming that the retention rate of each customer can be measured, this rate decides his/her CLV, and so the company needs focus on rising the retention rate, which increases the CLV, eventually profitability. Thus we now explain so-called retention marketing further below.

4.2 Retention Marketing

A key aspect of CRM is to focus on the existing customers rather than to attract new customers. That is, CRM tries to keep the existing customers make those

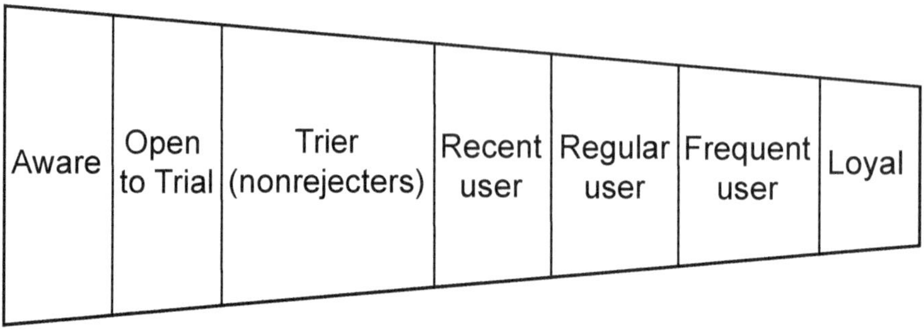

Figure 4.3: An example of marketing funnel: stages (from left to right) of changing customers in retention marketing.

customers more profitable for the company. Retention marketing is the main part of CRM for this purpose. This means that the choice of each individual customer would be monitored over time, resulting in a time-dependent model for customer behaviors [52].

4.2.1 Understanding Customers: Marketing Funnel

Fig. 4.3 shows one schematic example of stages (from left to right) of attracting and retaining customers, who are first just aware of the product and finally become the loyal customers of the product. This picture is called the *marketing funnel* or *customer journey mapping*. The marketing funnel in Fig. 4.3 has seven different categories (hereafter stages) of customers: Aware, Open to trial, Trier, Recent user, Regular user, Frequent user and Loyal, which are all ordered in time-series from left to right. Customers can be manually labeled by these stages from customer information or by a semi-manual way through data analysis, such as RFM analysis we already explained in Section 4.1.2. It would be easy to see, by the marketing funnel, through what types of stages strangers can be changed into loyal customers. There are two important points of retention marketing with marketing funnel.

Checking conversion rates: An usual analysis many companies can take would be that at some time point, the company can assign customers to each of the seven stages and check how many customers in each stage. However, more importantly, the company can check how many people are moved from one stage to the next stage. The ratio of customers which moved to the next stage to all customers in the previous stage is called the *conversion rate*, which can be rather simply computed as follows:

$$\text{Conversion rate} = \frac{N_{A\to B}^{(t)}}{N_A^{(t-1)}}, \tag{4.14}$$

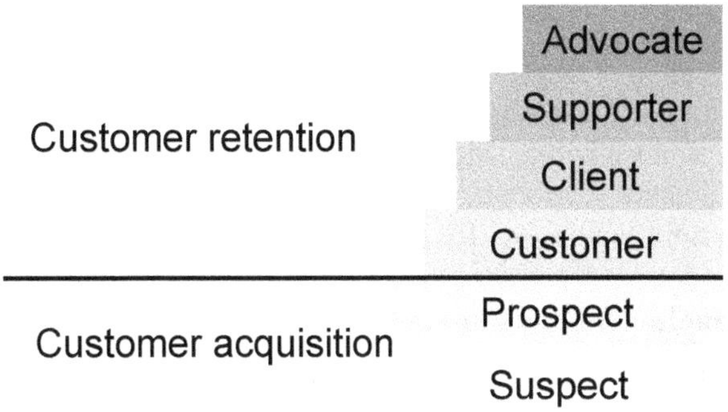

Figure 4.4: Another example of marketing funnel: stages (from bottom to top) of changing customers in retention marketing.

where $N_A^{(t-1)}$ is the number of customers at stage A at time $t-1$ and $N_{A\rightarrow B}^{(t)}$ is the number of customers who moved to stage B (which is subsequent to stage A) at time t.

By checking the conversion rates, the company may find the problem of each stage. For example, if the conversion rates for earlier stages (such as "Trier to Recent user" or "Regular user to Recent user") are low, the product has some significant, basic problem to be used repeatedly already.

Retention more than attraction: In general, for any product, attracting new customers is not easy. Thus instead of doing so, it would be rather easier to make customers retained and fostered to be loyal customers. In this sense, retention marketing should be more focused and paid attention. In summary, retaining existing customers is important to keep profitable customers for the company as well as, notably, avoid bad reputation of the goods from the customers.

Fig. 4.4 shows another example of the marketing funnel, in which stages are from the bottom to the top. Note that in this marketing funnel, all six stages use different terms from those in Fig. 4.3. In particular, the bottom two stages are regarded as those for customer acquisition, which are not necessarily customer retention. In fact several similar but slightly different marketing funnels, like Fig. 4.4, have been presented in the literature of marketing. However, the important points on retention marketing are totally common for all marketing funnels, even if their terms are not necessarily the same. Now the importance of retention marketing, i.e. retaining customers and boosting them to profitable customers, would be understood well, and then we will explain what can be done for promoting retention marketing below.

4.2.2 Boosting Customer Loyalty

The marketing funnel is helpful to understand customers of a given product, while the marketing funnel itself is unable to provide possible means the company can take. In this section we will briefly describe marketing techniques, which make the products appealing to customers more, to lead them to loyal customers. There are mainly two approaches for this purpose: direct and indirect approaches.

Direct Approach: Marketing Campaign Management

The marketing funnel is very useful to understand the current customers, while this would not tell what the company should do practically. A general approach for make customers move forward to the next stage of the marketing funnel is to execute campaigns of the products, such as discounting the price, bundle sales (cross-selling) and selling the product with some bonus. These campaigns need to be well designed and executed with a good timing so that the products are made attractive for customers. In fact if the campaigns are not managed well, the seller might even lose the customers, and so we can say that the direct approach has the danger of both the positive and negative results. Thus *marketing campaign management* contains all such aspects of providing products (or services) from the sellers.

Indirect Approach: Branding

Section 4.2.1 showed the way of understanding time-series behaviors of the customers on the product. Section 4.2.2 was on the campaign, which directly promotes the product, by which also customers can be promoted to the next stage of the marketing funnel.

Of course technically, as a baseline, companies should provide customers with better products and goods (or also services), which can exceed the expectation of customers. On the marketing side, however, an important point for increasing the loyal customers is *branding*. That is, the companies need make efforts to make the product, service and the company itself admitted as a *brand*, so that consumers can identify a particular brand, understanding the difference of the brands in the same product/service sector. The virtue (value) of the brand itself is called *brand equity*, which means the genuine effect or contribution of the brand only to the customer's preferences to the goods (or services), when taking the performance and quality of those goods aside.

Thus brands can be regarded as goods. Also the techniques to capture the reputation and evaluation of goods by customers can be used directly to understand the brand equity. For example, the way of building a competitive and distinguished brand is totally the same as market positioning of goods in Section 3.3. That is, readers can be referred to Section 3.3, regarding the goods as brands, by which product-customer relations can be replaced with brand-customer relations.

Table 4.2: Schematic example of a user-item matrix by users (rows) × items (columns).

	Item 1	Item 2	Item 3	Item 4	...
User 1					
User 2					
User 3					
User 4					
...					

4.3 Marketing Communications

We also briefly explain the ways which companies use to transmit the information on their company products to customers.

4.3.1 Types of Communication Channels

Traditionally it is said that between the sellers and buyers, there have existed at least the following eight communication channels [56]: 1) advertising, 2) sales promotion, 3) public relations and publicity, 4) events and experiences, 5) direct marketing, 6) interactive marketing, 7) word-of-mouth (WOM) marketing and 8) sales force. However in the modern internet era, the number of channels is now being increased and might be further so in the future. Dealing with these multiple channels will be discussed in Section 7.3.1.

4.3.2 Capturing Data through Communication Channels

By using such communications channels, we can have customer information, i.e. data on customers. Generally this is called *customer engagement*, and the analytics for customer engagement can change, depending on the stages of the marketing funnel ([13] is a review over customer analytics based on machine learning for analyzing customer engagement).

Practically the most fundamental data we can have for further analysis in marketing is so-called *user-item* matrices, where users indicate customers or buyers and items mean products, goods or services. Table 4.2 shows a schematic table of such a basic user-item matrix, in which rows are users and columns are items (which are reversible each other) and each entry is a binary record, meaning that the corresponding user bought the corresponding good. Note that once again this is the most basic data matrix in analysis of marketing, and also this matrix corresponds to the vector, i.e. the first data type introduced in Chapter 2, since each user is a vector for which each entry shows if the corresponding user purchased the corresponding item.

We will explain machine learning techniques, which can be run over this table in Chapters 5 to 7.

Chapter 5

Machine Learning for Marketing

We already introduced important terms and concepts in machine learning in Section 2.2. In this chapter, before we go into real applications of machine learning to problems of marketing, we explain general paradigms of machine learning and also rather simple approaches that can be used under those paradigms. These approaches can be directly used for applications in machine learning and also modified according to the change of the problem, model or assumption behind the model.

First of all, in both two major paradigms introduced in Section 2.2, i.e. supervised and unsupervised learning, the procedure of machine learning follows

$$\text{Data} \xrightarrow{\text{learning}} \text{Trained model} \xrightarrow{\text{prediction}} \text{Predicted result.}$$

That is, we first generate the framework of the model (with parameters to be trained) using some prior knowledge, and then by using given data, the model parameters are learned (trained), by which we can have the trained model. We then apply this trained model to the given unknown new data to have some result by prediction.

In this procedure, the first main step, i.e. learning or training the model, consists of the three aspects: 1) data, 2) model and 3) algorithm. Data was already explained in Section 2.1, in which we showed six data types, i.e. vectors, sets, sequences, trees, graphs and nodes in a graph, and also using multiple data types as input. We will, in this chapter, describe the other two aspects: models and algorithms.

There are various, diverse models already in machine learning, and selecting the most appropriate model out of them is an important issue, which is called *model selection*. Model selection is a discussion topic in any field to which machine learning methods are applied, and marketing is also so [90]. In reality the most proper model can be changed, depending on given data, the problem setting and also the assumption behind the setting. Thus, if we have no strong assumptions for the problem setting, we keep focusing on simple (or simplest) models in this book. For example, regarding supervised learning of vectors, we will consider linear regression all through this book, without considering any more complex model, like ensemble learning or neural networks. Again simply, regarding functions (such as loss functions), our focus is on simple polynomial functions, such as linear or quadratic ones, to represent the given data.

There are two reasons why we use such simple models:

1) The motivation of this book is to show machine learning approaches for problems of marketing. In particular, we will show the way of designing/modifying models and algorithms, depending on given data, problem and assumption behind the model. For this purpose we can start with a simple model, which can be modified into a more complex model, again according to the change in data, problem and assumption behind the model. Thus we just use a simple (usually the simplest) model under a given problem setting. For example, we focus on linear regression for supervised learning of vectors, instead of using some more complex models like deep learning (while we explain kernel learning in this chapter since kernelization is a possible modification for any case). Thus readers do not have to spend time for understanding a variety of models, while readers can replace the simplest model with some more complex one, if it is possible and readers are interested in doing so.

2) We use such simple models so that the optimization of estimating parameters of these models can be *convex* or *biconvex*, by which the optimum values of parameters can be obtained efficiently.

5.1 Organization of this Chapter

This chapter consists of three parts: 1) regular machine learning approaches, 2) feature learning and 3) kernel learning.

5.1.1 Regular Machine Learning

The first part shows typical (and the simplest) models and algorithms for a couple of data types. More in detail, in this part, we focus on two data types: 1) vectors and 2) nodes in a graph. The reason why we focus on these two types is:

Vectors: Vectors are most standard data in general, and also in marketing. For example, demographic data of individuals are vectors. A list of purchased goods for each user can be a vector (See Table 4.2). At the same time, for vectors, diverse models and learning algorithms have been already established and matured, more than other types of data. Thus from all three aspects of data, models and algorithms, vectors are the data to be focused on.

Nodes in a graph: An interesting data type we are facing in these years is networks, typically social networks. Then we pick up this data type, where the entire data is a graph, and each instance is a node in the graph. In social networks, nodes are individuals, where two nodes, connected by an edge, are thought to share some common property. More clearly, individuals linked in a social network share some common property like having the shared hobby, meaning that the edge between two nodes indicates the similarity between the connected two nodes. Thus examining nodes in a graph would be useful for understanding social networks and also exploring knowledge hidden in the networks.

For each of the two data types, we introduce both two major machine learning paradigms, i.e. supervised and unsupervised learning.

Vectors

In more detail, for vectors, we explain the following models and algorithms as machine learning examples.

Supervised learning: Linear regression Supervised learning is a typical setting in machine learning, and so even in marketing, we can raise interesting applications, such as forecasting customer demand on strongly weather dependepnt goods [107], measuring social media influencer index [6]. and so on. In general for superviesd learning, we can raise many models and algorithms, such as decision tree, support vector machine (SVM), deep learning etc. Among them, in this chapter, we show the simplest model and algorithm, i.e. linear regression, and two learning algorithms for this model. This is because linear regression is simple and a good example to understand supervised learning intuitively. Also algorithms for estimating parameters in linear regression are simple, and would be rather easy to understand, which also makes readers understand supervised learning intuitively more. Additionally, if maximizing the margin is used as an algorithm for estimating parameters in linear regression, the entire model and algorithm are equivalent to SVM, which uses the simplest kernel function, i.e. linear kernel. Thus it is also good to explain SVM by using linear regression.

Unsupervised learning: (constrained) K-means clustering Unsupervised learning for vectors can be divided mainly into two approaches: clustering and factorization. We show (constrained) K-means clustering as the most basic approach of clustering or unsupervised learning of vectors. We explain matrix factorization in Section 7.1.1 which is in fact well used for customer relationship marketing.

Nodes in a Graph

Also for nodes in a graph, we show a typical approach for each of the supervised and unsupervised learning. That is, the detail is below:

(Semi)supervised learning: label propagation Labels are assigned to nodes in a graph. In this setting labels are not for all nodes, and instead, part of nodes are labeled and the rest are not labeled. Then the problem is to assign labels to unlabeled nodes by using the labels of the nodes connected to the unlabeled nodes. This problem setting is called *semisupervised learning.* Also this setting of a graph is called *label propagation.*

Node clustering Unsupervised learning for nodes in a graph is to do clustering nodes in a graph by using the edge connections in the graph.

Thus we show four approaches, i.e. supervised and unsupervised learning approaches for each of the two data types: vectors and nodes in a graph.

Learning Multiple Data Types: Data Integrative Learning

We then further show supervised and unsupervised approaches for the case that multiple data types are given at the same time. In particular we focus on that as multiple data types, each instance can be represented by both a vector and also is a node in a graph. In practice, this means that for example, one instance is an individual who has both demographic data, representable by a vector, and can be a node in a social network. Thus, in this part of this chapter, we will show both supervised and unsupervised learning methods for each of the three different data types: 1) vectors, 2) nodes in a graph and 3) the combination of these two, which entirely result in six settings (supervised and unsupervised learning × three data types). We note that for each of the six settings, we raise at least one typical approach.

5.1.2 Feature Learning

Machine learning is mainly techniques to generate/extract patterns or hypothesis hidden in data. On the other hand, one aspect of machine learning is to modify the data itself. Thus in the second part of this chapter, we introduce this important aspect of machine learning which is called *feature learning*, particularly *feature extraction*, i.e. literally extracting important features out of given data. This technique is useful to reduce the redundancy of features in given data and/or summarize the information given by the data. In fact it would be always more useful if we have a larger number of instances, while a larger number of features are rather harmful, just causing redundancy more and also more time- and space-inefficiency. Thus it would be not only helpful but essential to select features most indispensable to the given data.

For feature learning, we focus on vectors as data, and consider the following two different approaches: 1) *feature selection*: which selects important features out of given data, and 2) *feature generation* or *dimensionality reduction*: which generates new, important features out of given data.

Note that we focus on machine learning-based general approaches, while in marketing features (variables) can be selected by more application-viewpoint, such as customer acquisition [100].

Feature Selection

For feature selection, we show two different approaches, depending on the two machine learning paradigms, i.e. supervised and unsupervised learning. In other words, unsupervised learning, feature selection can be performed by clustering features, in the sense that we can select only features, each representing each of the clusters over features. More in detail, we can select features, each being closest to each cluster center. That is, we do clustering over features (not instances) and then extract the most representative feature out of each cluster after clustering is complete.

In supervised learning, we can do clustering over features as the same manner in unsupervised learning, while we have label information, which should be used for feature learning as well. We then consider a general approach with two steps under the situation that given data has a huge number of features, such as a million of features. That is, this huge size of features need to be reduced to a moderate size by a rather simple method first, and then this set of features can be reduced further by a more careful method.

Feature Generation/ Dimensionality Reduction

On the other hand, for feature generation, we show the most typical dimensionality reduction method, principal component analysis (PCA), which generates new, most symbolic features out of given data, without using labels. Thus in the sense that PCA does not use any label information, PCA is one approach in unsupervised learning.

5.1.3 Kernel Learning

We then finally, in the third (final) part of this chapter, show one type of machine learning method, called *kernel learning* or *kernel method.* In kernel learning, the input is represented by a kernel function, which is given by users to define the similarity between two given instances. which can incorporate prior/ background knowledge of the application. Thus kernel learning can be applied to any data, regardless of data types, such as vectors, nodes over a graph or combination of them, etc. Also kernel learning uses a kernel function as input instead of real given data. Thus kernel learning can cover a variety of machine learning paradigms, such as supervised and unsupervised learning and feature learning. We explain the idea, common procedure and sample machine learning methods, of kernel learning, for each of supervised, unsupervised and feature learning.

5.2 Regular Machine Learning

5.2.1 Learning Vectors

Let $\boldsymbol{X}$ be a given set of instances, i.e. $(N \times M)$-matrix, where the i-th row, $\boldsymbol{x}_i^\top$, is the i-th instance of all N instances, and each column is one of M features. Let $\boldsymbol{y}$ be a given vector of labels, where the size of $\boldsymbol{y}$ is N.

Supervised Learning

The idea of supervised learning is to predict the label by using given features. Mathematically this is equivalent to estimate the following function f, given $\boldsymbol{y}$ and $\boldsymbol{X}$.

$$\boldsymbol{y} \sim f(\boldsymbol{X}) \tag{5.1}$$

Model Example: Linear Regression

As one of the simplest models of supervised learning, we consider linear regression, which implements the idea of approximating $\boldsymbol{y}$ by using linear combination[1] of features in $\boldsymbol{X}$. Letting x_{ij} be the (i,j)-element of matrix $\boldsymbol{X}$, and y_i be the i-th element of $\boldsymbol{y}$. Linear regression can be written as follows:

$$\begin{pmatrix} y_1 \\ y_2 \\ \vdots \\ y_N \end{pmatrix} \sim \beta_1 \begin{pmatrix} x_{11} \\ x_{21} \\ \vdots \\ x_{N1} \end{pmatrix} + \beta_2 \begin{pmatrix} x_{12} \\ x_{22} \\ \vdots \\ x_{N2} \end{pmatrix} + \cdots + \beta_M \begin{pmatrix} x_{1M} \\ x_{2M} \\ \vdots \\ x_{NM} \end{pmatrix} + \begin{pmatrix} \beta_{01} \\ \beta_{02} \\ \vdots \\ \beta_{0N} \end{pmatrix}, \tag{5.2}$$

where $\beta_1, \ldots, \beta_M$ and $\beta_{01}, \ldots, \beta_{0N}$ are real-valued coefficients or parameters to be estimated from both $\boldsymbol{X}$ and $\boldsymbol{y}$.

For example, if we have only two features, this model can be written as follows:

$$\begin{pmatrix} y_1 \\ y_2 \\ \vdots \\ y_N \end{pmatrix} \sim \beta_1 \begin{pmatrix} x_{11} \\ x_{21} \\ \vdots \\ x_{N1} \end{pmatrix} + \beta_2 \begin{pmatrix} x_{12} \\ x_{22} \\ \vdots \\ x_{N2} \end{pmatrix} + \begin{pmatrix} \beta_{01} \\ \beta_{02} \\ \vdots \\ \beta_{0N} \end{pmatrix}, \tag{5.3}$$

where again β_1, β_2 and $\beta_{01}, \ldots, \beta_{0N}$ are real-valued coefficients or parameters to be estimated from both $\boldsymbol{X}$ and $\boldsymbol{y}$.

In marketing, linear regression has been used for various purposes, such as finding key product attributes for consumer's purchase intention [9].

In this case, the linear regression just considers two-dimensional (2D) space, which can be generated by the two features. Fig. 5.1 shows an example of such

[1]The linear combination of features means that weighted features are used to predict $\boldsymbol{y}$. In fact a simpler approach would be no weights over features to predict $\boldsymbol{y}$. This approach is called a *naive Bayes classifier*. In marketing, for example, this approach is used to forecast product sales from online reviews and historical sales data [32]. Interested readers can refer to [70] for more details on supervised learning.

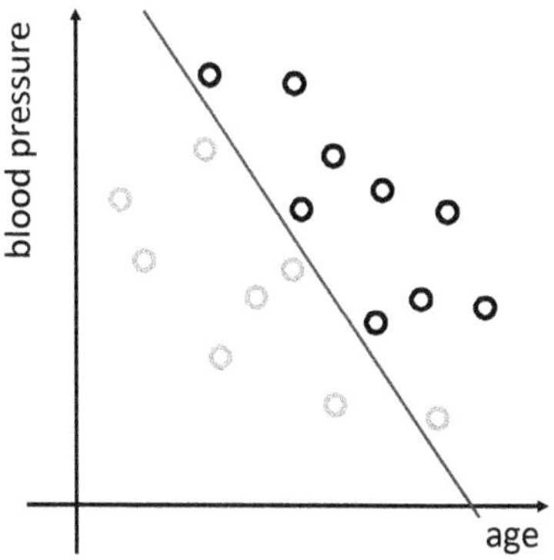

Figure 5.1: Example of linear regression over a two-dimensional space made by two features.

a 2D space, being made of two features. We can see that linear regression is equivalent to divide the space of arbitrary dimensions (defined by the features) into, for example, two subspaces, like Fig. 5.1.

We can then see that we can write both (5.2) and (5.3) into the following matrix form:

$$\boldsymbol{y} \sim \boldsymbol{X}\boldsymbol{\beta} + \boldsymbol{b}. \tag{5.4}$$

As mentioned in (5.1), supervised learning is to estimate the function f, and in linear regression this f is assumed to be a linear function as follows:

$$f(\boldsymbol{X}) = \boldsymbol{X}\boldsymbol{\beta} + \boldsymbol{b}. \tag{5.5}$$

Furthermore before trying to estimate parameters $\boldsymbol{\beta}$ and $\boldsymbol{b}$ from given data $\boldsymbol{X}$, we can normalize $\boldsymbol{X}$ so that $\boldsymbol{b}$ becomes a zero vector (all elements are zero). By doing so, we can see linear regression is the following simpler function:

$$f(\boldsymbol{X}) = \boldsymbol{X}\boldsymbol{\beta}. \tag{5.6}$$

Also the model can be as follows:

$$\boldsymbol{y} \sim \boldsymbol{X}\boldsymbol{\beta}. \tag{5.7}$$

Below we raise two algorithms for estimating $\boldsymbol{\beta}$, based on different ideas: 1) method of least squares and 2) maximizing the margin.

Algorithm for Linear Regression: 1) Method of Least Squares

The idea of this algorithm is to minimize the least square error[2] between the right- and left-hand sides of (5.7), for all instances. This means that the estimated label

[2]The squared distance (error) is well used in machine learning. See Section 5.5.1 for more explanation on this point.

of each instance should be consistent with the corresponding true label, for all instances. This can be shown as the following optimization problem:

$$\text{Minimize} \quad \epsilon^2, \quad \text{where } \epsilon = ||\boldsymbol{y} - \boldsymbol{X}\boldsymbol{\beta}|| \tag{5.8}$$

This problem[3] can be solved by taking the partial derivative of the following Lagrangian and setting that to zero with respect to $\boldsymbol{\beta}$:

$$L(\boldsymbol{\beta}) \quad = \quad ||\boldsymbol{y} - \boldsymbol{X}\boldsymbol{\beta}||^2 \tag{5.9}$$

This results in the following analytical solution to estimate $\boldsymbol{\beta}$, i.e. $\hat{\boldsymbol{\beta}}$ (see [70] for derivation):

$$\hat{\boldsymbol{\beta}} = (\boldsymbol{X}^\top \boldsymbol{X})^{-1} \boldsymbol{X}^\top \boldsymbol{y} \tag{5.10}$$

However, (5.10) has two problems:

Nonsingular matrix: (5.10) requests that $\boldsymbol{X}^\top\boldsymbol{X}$ is always a nonsingular (invertible) matrix, while we are unable to guarantee that $\boldsymbol{X}^\top\boldsymbol{X}$ is nonsingular, because $\boldsymbol{X}$ is given data.

Overfitting: Also in (5.10), $\hat{\boldsymbol{\beta}}$ is obtained from $\boldsymbol{X}$ only, implying that $\hat{\boldsymbol{\beta}}$ is likely to overfit the given data $\boldsymbol{X}$ only, lacking the generalization of $\boldsymbol{\beta}$.

Then to solve the above two problems at the same time, we place constraints over $\boldsymbol{\beta}$. A standard constraint is the following, so-called L^p norm:

$$||\boldsymbol{\beta}||_p^p = (\sum_i \beta_{0i}^p)^{\frac{1}{p}} = (\beta_{01}^p + \beta_{02}^p + \cdots + \beta_{0N}^p)^{\frac{1}{p}} = C, \tag{5.11}$$

where C is a constant[4].

We can use the L^p norm ($C = 1$) as a constraint of (5.9), and the new Lagrangian can be written as follows:

$$L(\boldsymbol{\beta}) \quad = \quad ||\boldsymbol{y} - \boldsymbol{X}\boldsymbol{\beta}||^2 + \lambda(||\boldsymbol{\beta}||^p - 1), \tag{5.13}$$

where the constraint term $\lambda(||\boldsymbol{\beta}||^p - 1)$ is called a *regularizer*, and its coefficient, i.e. λ, is a hyperparameter[5], called *regularization coefficient*, one type of hyperparameters. In fact, the constraint can be changed by changing p. Fig. 5.2 shows the possible range of values to be taken by parameters under the constraint of L^p norm. For example, in (a), i.e. p=0, this constraint (regularizer) is zero, and the new Lagrangian is kept the same as the original Lagrangian of (5.9). On the other hand, for (b) $p = 1$ or (c) $p = 2$, the possible values of the parameters are on the linear lines or on the circle, respectively. That's why these terms are called regularizers.

---

[3] Just in case, $||\cdot||$ means the Euclidean norm (and equivalently L^p norm and $p = 2$). For example, for $\boldsymbol{a}^\top = (a_1, \ldots, a_N)$, $||\boldsymbol{a}|| = \sqrt{\sum_i a_i^2}$

[4] We often fix $C = 1$, and this case L^p norm can be written as follows:

$$1 = (\sum_i \beta_{0i}^p)^{\frac{1}{p}} = \sum_i \beta_{0i}^p = ||\boldsymbol{\beta}||_1^p \tag{5.12}$$

Furthermore $||\boldsymbol{\beta}||_1^p$ is often just written as $||\boldsymbol{\beta}||^p$.

[5] A hyperparameter is a parameter, which is not estimated by the process of machine learning but specified by a user or empirically determined.

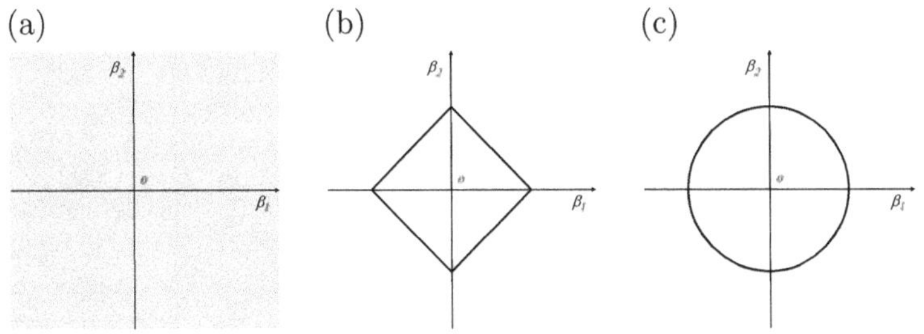

Figure 5.2: Possible parameter values taken when (a) $p = 0$, (b) $p = 1$, and (c) $p = 2$.

Procedure 5.1: RIDGE REGRESSION.

Input: matrix (vectors): $\boldsymbol{X}$; label: $\boldsymbol{y}$; regularization coefficient: λ
Output: linear coefficients: $\boldsymbol{\beta}$

Estimate $\hat{\boldsymbol{\beta}}$ through (5.15) from $\boldsymbol{X}$, $\boldsymbol{y}$ and λ.;
Output finally estimated $\hat{\boldsymbol{\beta}}$.;

Thus for example, for $p = 2$, the Lagrangian can be written as follows:

$$L(\boldsymbol{\beta}) \quad = \quad ||\boldsymbol{y} - \boldsymbol{X}\boldsymbol{\beta}||^2 + \lambda(||\boldsymbol{\beta}||^2 - 1) \tag{5.14}$$

Taking the partial derivative of (5.14) and setting that to zero with respect to $\boldsymbol{\beta}$, we can have the following analytical solution again (see [70] for derivation):

$$\hat{\boldsymbol{\beta}} = (\boldsymbol{X}^\top \boldsymbol{X} + \lambda \boldsymbol{I})^{-1} \boldsymbol{X}^\top \boldsymbol{y}, \tag{5.15}$$

where $\boldsymbol{I}$ is the identity matrix. This parameter estimation with L^2 norm is called *ridge regression*. **Procedure 5.1** shows a pseudocode of this algorithm. Then once β is estimated as $\hat{\boldsymbol{\beta}}$, we can use this parameter to estimate the unknown label of given new instance $\boldsymbol{x}_{\text{new}}$, as follows:

$$y_{\text{new}} = \boldsymbol{x}_{\text{new}}^\top \hat{\boldsymbol{\beta}} \tag{5.16}$$

In fact the performance of ridge regression is already confirmed in marketing applications [69]. A further advantage of this estimation is that by keeping λ as a certain large value, the term, $\boldsymbol{X}^\top \boldsymbol{X} + \lambda \boldsymbol{I}$, can be a nonsingular matrix, by which the inverse of this matrix can be always obtained. Also another good point of this result is that overfitting $\boldsymbol{X}$ can be avoided at the same time. Thus the two problems of linear regression without any regularization terms are solved by this formulation.

However one item of note is that if λ is extremely large, the term $\boldsymbol{X}^\top \boldsymbol{X} + \lambda \boldsymbol{I}$ is just approximated by $\boldsymbol{I}$, meaning that this term becomes nothing to do with

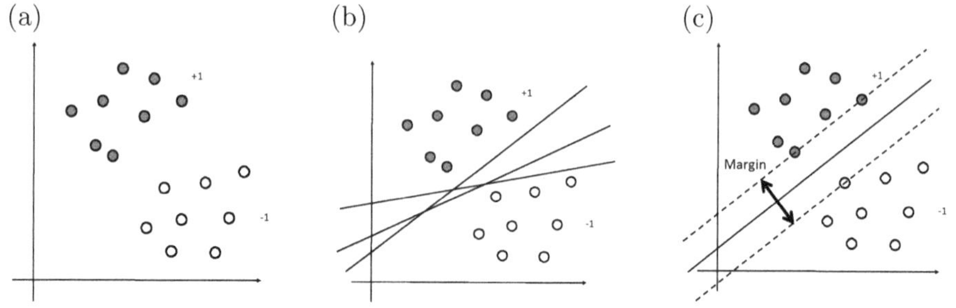

Figure 5.3: Margin: key of support vector machine (from [70]).

given data $\boldsymbol{X}$. Thus, although the above formulation looks good by avoiding the nonsingular matrix and also overfitting, clearly this way is not necessarily the best way to do so. Thus this would be a good motivation to think about another algorithm (based on margin maximization), which will be explained below.

Back to L^p norm, linear regression with L^1 norm is called *Least Absolute Shrinkage and Selection Operator* (LASSO), where as shown in Fig. 5.2 (b), parameter values are on the lines. The point of LASSO is that parameter values cannot be equally taken on the lines in Fig. 5.2 (b), and rather are likely to be on the intersections of lines, meaning that many parameter values are forced to be zero. That is, the solutions obtained by LASSO, more generally L^1 norm, are likely to have many zeros and only a small number of non-zero. These solutions are called *sparse* solution, which are useful for understanding the data or interpretation of obtained results, and this type of learning is called *sparse learning.* Thus, when the input is vectors, sparse learning has connection to feature learning, which will be explained in Section 5.1.2.

Also regarding L^p norm, we can take another approach of both using L^2 norm and L^1 norm as the regularizer as follows:

$$L(\boldsymbol{\beta}) = ||\boldsymbol{y} - \boldsymbol{X}\boldsymbol{\beta}||^2 + \lambda_2(||\boldsymbol{\beta}||^2 - 1) + \lambda_1(||\boldsymbol{\beta}|| - 1) \tag{5.17}$$

This formulation is called *elastic net.*

Algorithm for Linear Regression: 2. Maximizing the Margin

The idea of the method of least squares was to minimize the error between the estimated label and the true label, for all instances. However, for example, in binary classification, as seen by Fig. 5.1, if some (linear) function can discriminate the instances distributed around the boundary of the function, the other instances, say those distributed far from the linear function, can be automatically discriminated.

This means that, we can first focus just on the instances which are located in the boundary of the classification/regression function. For example, Fig. 5.3 (a) shows the distributions of instances, consisting of two classes. Then, like (b) in Fig. 5.3, the problem becomes of whether we place a linear function between these two classes. We can then easily see that the ideal linear function for our situation of discriminating the two classes is the function which maximizes the the distance between the function and the closest instance (from each class). This distance is called the *margin*. That is, the linear function should be placed just between the two instances, each being the closest instance from each of the two classes. Fig. 5.3 (c) shows exactly this situation. This idea of placing the function so that the distance between the function and the closest instance from one class is maximized is called *margin maximization* or *maximizing the margin*. In reality the given data usually has a number of features, which makes the space by features high-dimensional, and this space is called a *(high-dimensional) feature space*, and the function of discriminating between two classes is called a *hyperplane*. Also those instances closest to the hyperplane are called *support vectors*.

More in detail, for simplicity, on the two-dimensional space, we can first write the linear function as follows:

$$y = \boldsymbol{\beta}^{\mathsf{T}}\boldsymbol{x} + b. \tag{5.18}$$

Let $(\boldsymbol{x}_{+}, +1)$ and $(\boldsymbol{x}_{-}, -1)$ be two support vectors of the two classes, say positives and negatives, respectively. Also let $(\boldsymbol{x}_0, 0)$ be the point on the ideal linear function, which satisfies the idea of margin maximization.

Then due to the idea of maximizing the margin, the following three must be kept:

$$+1 = \boldsymbol{\beta}^{\mathsf{T}}\boldsymbol{x}_{+} + b. \tag{5.19}$$

$$0 = \boldsymbol{\beta}^{\mathsf{T}}\boldsymbol{x}_{0} + b. \tag{5.20}$$

$$-1 = \boldsymbol{\beta}^{\mathsf{T}}\boldsymbol{x}_{-} + b. \tag{5.21}$$

From the difference of the two support vectors $(\boldsymbol{x}_{+}, +1)$ and $(\boldsymbol{x}_{-}, -1)$ of the above, we can have the following:

$$+1 - (-1) = \boldsymbol{\beta}^{\mathsf{T}}(\boldsymbol{x}_{+} - \boldsymbol{x}_{-}). \tag{5.22}$$

In the above, $(\boldsymbol{x}_{+} - \boldsymbol{x}_{-})$ is exactly the margin, which should be maximized, while the left-hand side is just a constant. This means that to maximize the margin, we can minimize the absolute value of $\boldsymbol{\beta}$, say $||\boldsymbol{\beta}||$ (or more generally $||\boldsymbol{\beta}||^p$, where p is an arbitrary value).

So far we have considered the instances of boundaries (or support vectors), while obviously all instances have to keep the following constraints:

$$\boldsymbol{\beta}^{\mathsf{T}}\boldsymbol{x} + b \geq +1 \quad \text{for} \quad y \geq 1. \tag{5.23}$$

$$\boldsymbol{\beta}^{\mathsf{T}}\boldsymbol{x} + b < -1 \quad \text{for} \quad y < -1. \tag{5.24}$$

This can be written in the following way:

$$y_i(\boldsymbol{\beta}^{\mathsf{T}}\boldsymbol{x}_i + b) \geq 1 \quad \text{for all} \quad i. \tag{5.25}$$

Overall the margin maximization criterion can be formulated into the following optimization problem.

$$\min_{\boldsymbol{w}} \frac{1}{2}||\boldsymbol{\beta}||^2 \quad \text{subject to} \quad y_i(\boldsymbol{\beta}^\mathsf{T}\boldsymbol{x}_i + b) \geq 1 \quad \text{for all} \quad i. \tag{5.26}$$

That is, this problem can be solved by using the following Lagrangian:

$$L(\boldsymbol{\beta}, b, \boldsymbol{\alpha}) = \frac{1}{2}\boldsymbol{\beta}^\mathsf{T}\boldsymbol{\beta} - \sum_i \alpha_i(y_i(\boldsymbol{\beta}^\mathsf{T}\boldsymbol{x}_i + b) - 1), \tag{5.27}$$

where $\alpha_i (i = 1, \ldots, N)$ are Lagrange multipliers.

This is the derivation of the case in which instances can be discriminated clearly in given data (which is called *hard margin*), while in reality, the instances cannot be so easily separated by the hyperplane and this point can be considered in the optimization (now the margin is called *soft margin*). We note that the derivation for soft margin is substantially equivalent to that of the above hard margin. More importantly the formulation of the above Lagrangian of soft margin is kept as the same as hard margin essentially[6].

The description of learning linear regression based on margin maximization continues to support vector machine in kernel learning in Section 5.4.3.

Linear Regression Summary

We first note that there exist (two) different algorithms for linear regression, i.e. the same model, at the level of Lagrangian in optimization, as follows:

Method of least squares:

$$L(\boldsymbol{\beta}, \lambda) = ||\boldsymbol{y} - \boldsymbol{X}\boldsymbol{\beta}||^2 + \lambda(||\boldsymbol{\beta}||^p - 1). \tag{5.28}$$

Margin maximization:

$$L(\boldsymbol{\beta}, b, \boldsymbol{\alpha}) = \frac{1}{2}\boldsymbol{\beta}^\mathsf{T}\boldsymbol{\beta} - \sum_i \alpha_i(y_i(\boldsymbol{\beta}^\mathsf{T}\boldsymbol{x}_i + b) - 1). \tag{5.29}$$

That is, we can build different algorithms based on different ideas/motivations for the same model.

We just have showed a simple example, linear regression and its learning algorithms, while note that there are many models and algorithms for supervised learning of vectors, including decision tree, logistic regression, deep learning, etc.

Unsupervised Learning

The other major paradigm of machine learning is unsupervised learning, where we do not have labels, i.e. $\boldsymbol{y}$, and then only matrix $\boldsymbol{X}$ is given. The idea of unsupervised learning is to summarize the data, and typical unsupervised learning

[6] For more detailed derivation of both hard and soft margins, and also further derivation of optimization, readers can refer to [70].

is clustering and factorization. We here consider clustering, particularly the most standard clustering method, called (constrained) K-means clustering. In fact another typical clustering method is based on probabilistic modeling, e.g. topic models or latend Dirichlet allocation (LDA), which have been used in marketing for text data [80, 4]. Factorization will be described from the scratch in Section 7.1.1, since factorization is related with relationship marketing more than target marketing.

Model: K-means Clustering

Clustering is one typical approach for summarizing data, i.e. unsupervised learning, and in clustering, given instances are summarized into several groups, which we call *clusters*.

We now think about the situation, in which clustering of given data is already done and we have a new instance, which should be assigned to one of the clusters already generated from given data. Under this situation, to choose the optimal cluster, into which the new instance should be fallen, the most reasonable approach would be first to check the distance between this instance and the representative instance of each cluster and second to select the cluster for which the distance between the new instance and the representative instance is smallest. In fact this idea can be applied to do clustering given training data, assuming that the way of computing the representative instances of all clusters are already fixed.

Let $\boldsymbol{\mu}_k$ be the representative instance (vector) in the k-th cluster. The above idea can be formulated as the minimization of the following term, in terms of $\boldsymbol{Z}$ (see below on $\boldsymbol{Z}$):

$$\sum_{k=1}^{K}\sum_{i=1}^{N} \boldsymbol{Z}_{ik}\mathrm{Dist}(\boldsymbol{x}_i, \boldsymbol{\mu}_k), \tag{5.30}$$

where $\mathrm{Dist}(\boldsymbol{a}, \boldsymbol{b})$ is the distance between two vectors $\boldsymbol{a}$ and $\boldsymbol{b}$, and $\boldsymbol{Z}$ is a binary cluster assignment matrix, in which for the i-th instance $\boldsymbol{x}_i$, the following must be kept:

$$\boldsymbol{Z}_{ik'} = 1 \qquad \text{if} \qquad k' = \arg\min_k \mathrm{Dist}(\boldsymbol{x}_i, \boldsymbol{\mu}_k), \tag{5.31}$$

$$\boldsymbol{Z}_{ik} = 0 \quad \text{otherwise}. \tag{5.32}$$

There would be many ways for the representative instance $\boldsymbol{\mu}_k$ of cluster k, while statistically a reasonable choice for the representative vector of one cluster is the *mean* over all instances in the corresponding cluster, as shown below. Then the clustering with this idea is called *K-means clustering*, where again K means the number of clusters.

$$\boldsymbol{\mu}_k \quad \leftarrow \quad \frac{\sum_{i|\boldsymbol{x}_i \in C_k} \boldsymbol{x}_i}{|C_k|}, \tag{5.33}$$

where C_k is the set of instances in the k-th cluster, $\boldsymbol{x} \in C_k$ is instance $\boldsymbol{x}$ in cluster C_k and $|C_k|$ is the number of instances in cluster C_k.

The K of K-means clustering indicates the number of clusters which is specified before starting clustering. The learning process of K-means clustering is to estimate cluster assignment matrix $\boldsymbol{Z}$ and also the representative instance, i.e. the mean, $\boldsymbol{\mu}$, of each cluster. That is, we need to estimate both $\boldsymbol{Z}$ and mean $\boldsymbol{\mu}$ at the same time.

For more detailed description and related work on clustering, see Chapter 3 of [70].

Algorithm: Constrained K-means Clustering

The model description on K-means clustering indicates that each instance $\boldsymbol{x}$ should be assigned to the cluster which has the mean closest to $\boldsymbol{x}$. This idea behind K-means clustering can be formulated as the following optimization problem:

$$\min_{\boldsymbol{Z},\boldsymbol{\mu}} \sum_{k=1}^{K} \sum_{i=1}^{N} Z_{ik} \mathrm{Dist}(\boldsymbol{x}_i, \boldsymbol{\mu}_k). \tag{5.34}$$

Again statistically a reasonable choice for the representative vector of a cluster is the mean over all instances in the corresponding cluster. Then a further reasonable choice for the distribution of instance $\boldsymbol{x}_i$ in cluster k (its representative vector is $\boldsymbol{\mu}_k$) would be the normal distribution:

$$C \exp(\frac{||\boldsymbol{x}_i - \boldsymbol{\mu}_k||^2}{C'}), \tag{5.35}$$

where C and C' are both functions with *variance* as their only parameter: $C = \sqrt{2\pi}\sigma$ and $C' = \sigma^2$, where σ is the variance. This leads to the squared distance for $\mathrm{Dist}(\boldsymbol{x}_i, \boldsymbol{\mu}_k)$ as again a reasonable choice, meaning that the sum over the distance between each instance and the mean would be proportional to the variance. That is, the variance in this case is equivalent to the squared distance for $\mathrm{Dist}(\boldsymbol{x}_i, \boldsymbol{\mu}_k)$, as follows:

$$\mathrm{Dist}(\boldsymbol{x}_i, \boldsymbol{\mu}_k) \quad = \quad ||\boldsymbol{x}_i - \boldsymbol{\mu}_k||^2, \tag{5.36}$$

then (5.34) can be written as follows:

$$\min_{\boldsymbol{Z},\boldsymbol{\mu}} \sum_{k=1}^{K} \sum_{i=1}^{N} \boldsymbol{Z}_{ik} ||\boldsymbol{x}_i - \boldsymbol{\mu}_k||^2. \tag{5.37}$$

By using the nature of the variance (or squared distance), this objective function (5.37) can be transformed into the following (see Chapter 3 of [70] for this transformation):

$$\sum_{k=1}^{K} \sum_{i=1}^{N} \boldsymbol{Z}_{ik} ||\boldsymbol{x}_i - \boldsymbol{\mu}_k||^2 \quad = \quad \sum_{k=1}^{K} \sum_{j=1}^{N} \sum_{i=1}^{N} \boldsymbol{Z}_{ik} \boldsymbol{E}_{ij} \boldsymbol{Z}_{jk} \tag{5.38}$$

$$= \quad \mathrm{trace}(\boldsymbol{Z}^{\mathsf{T}} \boldsymbol{E} \boldsymbol{Z}), \tag{5.39}$$

where the element of the i-th row and the j-th column of $\boldsymbol{E}$, i.e. $\boldsymbol{E}_{ij}$, can be given as follows:

$$\boldsymbol{E}_{ij} = \frac{1}{2N}||\boldsymbol{x}_i - \boldsymbol{x}_j||^2. \tag{5.40}$$

Furthermore we can focus on binary cluster assignment matrix $\boldsymbol{Z}$, considering the nature of $\boldsymbol{Z}$: each instance can be assigned into only one cluster out of all K clusters. That is, for each instance (one row), only one element has one and other elements are zero. This means that when we pick up any two column vectors, these two vectors do not have one for the same row, indicating that the inner product of arbitrary two column vectors, say $\boldsymbol{z}_k$ and $\boldsymbol{z}_{k'}$, becomes zero, unless $k = k'$, and also if $k = k'$, this inner product is equivalent to the size (number of instances) in the k-th cluster, i.e. $|C_k|$. This can be written in the following condition:

$$\boldsymbol{z}_k^{\top}\boldsymbol{z}_{k'} = 0 \quad \text{if} \quad k \neq k', \tag{5.41}$$
$$\boldsymbol{z}_k^{\top}\boldsymbol{z}_k = |C_k| \quad \text{otherwise.} \tag{5.42}$$

This can be written in a matrix form as follows:

$$\boldsymbol{Z}^{\top}\boldsymbol{Z} = \begin{pmatrix} |C_1| & 0 & \cdots & 0 \\ 0 & |C_2| & 0 & \vdots \\ \vdots & 0 & \ddots & 0 \\ 0 & \cdots & 0 & |C_K| \end{pmatrix}. \tag{5.43}$$

That is, we can write this in a matrix form:

$$\boldsymbol{Z}^{\top}\boldsymbol{Z} = \boldsymbol{D}, \tag{5.44}$$

where $\boldsymbol{D}$ is the diagonal matrix with each diagonal element specifying the size (number of instances) of the corresponding cluster.

If we follow the idea that the cluster size should be balanced (or the cluster size should be kept the same over all clusters), i.e. $|C_1| = |C_2| = \cdots = |C_k| = |C|$, (5.43) can be written as follows:

$$\boldsymbol{Z}^{\top}\boldsymbol{Z} = \lambda \boldsymbol{I}, \tag{5.45}$$

where $\boldsymbol{I}$ is the identity matrix and λ is the size of each cluster.

We can specify the sizes of K clusters at arbitrary values, by using the constraint shown in (5.44), where again $\boldsymbol{D}$ is a diagonal matrix, in which each diagonal element is set as the size of each cluster which we want to specify.

We can then set up this as a constraint for optimizing the cluster assignment matrix of K-means clustering. Over all the optimization problem can be formulated as follows:

$$\min_{\boldsymbol{Z}} \quad \text{trace}(\boldsymbol{Z}^{\top}\boldsymbol{E}\boldsymbol{Z}) \text{ subject to } \boldsymbol{Z}^{\top}\boldsymbol{Z} = \boldsymbol{D}. \tag{5.46}$$

Procedure 5.2: CLUSTERING POSTPROCESSING.

Function `ClusteringPostProcessing(`$\boldsymbol{Z}$`)`
  Input: cluster assignment matrix: $\boldsymbol{Z}$
  Output: cluster assignment matrix: $\boldsymbol{Z}$
  repeat
    foreach *cluster* k **do** cluster center estimation
      Estimate center $\boldsymbol{\mu}_k$ using instances by (5.33).;
    foreach *instance* i **do** instance assignment to cluster
      foreach *cluster* k **do**
        Compute distance between instance i and cluster k, $\mathrm{Dist}(\boldsymbol{x_i}, \boldsymbol{\mu}_k)$, by using (5.36).;
      Find the closest cluster $\hat{k} \leftarrow \arg\min_k \mathrm{Dist}(\boldsymbol{x_i}, \boldsymbol{\mu}_k)$.;
      Update $\boldsymbol{Z}_{ik} = 1$ if $k = \hat{k}$; otherwise $\boldsymbol{Z}_{ik} = 0$.;
  until *convergence*;
  Output finally estimated $\hat{\boldsymbol{Z}}$.;

By relaxing the binary constraint of the cluster assignment matrix $\boldsymbol{Z}$ so that $\boldsymbol{Z}$ can have real-valued elements, then we can obtain this matrix through an optimization problem. The Lagrangian of this optimization algorithm can be written as follows:

$$L(\boldsymbol{Z}, \lambda) = \mathrm{trace}(\boldsymbol{Z}^\top \boldsymbol{E} \boldsymbol{Z}) - \lambda\, \mathrm{trace}(\boldsymbol{Z}^\top \boldsymbol{Z} - \boldsymbol{D}). \tag{5.47}$$

By setting the partial derivative of this Lagrangian with respect to $\boldsymbol{Z}$ to zero, we can have the following eigenvalue problem:

$$\boldsymbol{E}\boldsymbol{Z} = \lambda \boldsymbol{Z}. \tag{5.48}$$

Thus after solving the eigenvalue problem and also after a postprocessing step, we can have the estimated binary cluster assignment matrix $\hat{\boldsymbol{Z}}$.

The postprocessing is an alternate procedure of the two steps, by using, for each instance, the corresponding row of $\boldsymbol{Z}$ (instead of $\boldsymbol{X}$), as follows:

1. **Cluster center estimation** For each cluster, we estimate the center from instances assigned to the cluster, by using (5.33) (Note again that each instance is not $\boldsymbol{x}_i$ but the i-th row of $\boldsymbol{Z}$.).

2. **Instance assignment to cluster** We assign each instance to one cluster where the cluster should be selected so that the instance should be assigned to the closest cluster in terms of the distance between the instance and the cluster center.

We repeat the above alternating postprocessing two steps until convergence, and then the resultant $\boldsymbol{Z}$ has the final cluster assignment of (constrained) K-means clustering. **Procedure 5.2** summarizes the above postprocessing procedure as a

Procedure 5.3: K-MEANS CLUSTERING.

```
Function K-means Clustering(X)
    Input: Matrix (instances × features): X
    Output: Cluster assignment matrix: Z
    Compute E from X.;
    Estimate Ẑ by solving the eigenvalue problem given in (5.48).;
    Run ClusteringPostProcessing(Ẑ).;
    Output finally estimated Ẑ.;
```

pseudocode. **Procedure 5.3** shows the entire procedure of K-means clustering as a pseudocode. For more detailed derivation and explanation on the constrained K-means clustering algorithm, see Chapter 3 of [70].

5.2.2 Learning Nodes in a Graph

The setting of "nodes in a graph" assumes the graph with unique nodes, meaning that all nodes are different. This setting is reasonable, as mentioned in Section 2.1.6, by considering possible applications, such as social networks and gene regulatory network, where each node is an individual or a gene. Fig. 5.4 shows a toy example with eight nodes, indicating that the graph with unique nodes corresponds to an *adjacency matrix*, in which both rows and columns have the same size, corresponding to the number of nodes, meaning that both rows and columns are instances.

Let $\boldsymbol{G}$ be a given graph and $\boldsymbol{W}$ be the adjacency matrix, corresponding to $\boldsymbol{G}$. That is, if two nodes in $\boldsymbol{G}$ are connected by an edge, the corresponding element of $\boldsymbol{W}$ is 1; otherwise zero. Also let $\boldsymbol{D}$ be the diagonal matrix, in which diagonal $\boldsymbol{D}_{ii}$ is given by

$$\boldsymbol{D}_{ii} = \sum_j \boldsymbol{W}_{ij} \tag{5.49}$$

We call $\boldsymbol{L}$ $(= \boldsymbol{D} - \boldsymbol{W})$ *graph Laplacian* of $\boldsymbol{G}$.

Graph Smoothness

Given a graph as data, a question is the meaning behind each edge between two nodes. We can think that the two nodes connected by an edge should be similar. This would be reasonable, since for example, in social networks, when two nodes corresponding to two individuals should share some common property, such as a kind of interest. Also in gene regulatory network, two genes (nodes) connected by an edge should share the same gene function or more generally some biological mechanism. Thus, when some values or labels are assigned to the two nodes, which are connected by an edge, these values should be similar. That is, node labels (or assigned values) should be consistent if these nodes are connected by

(a) Graph

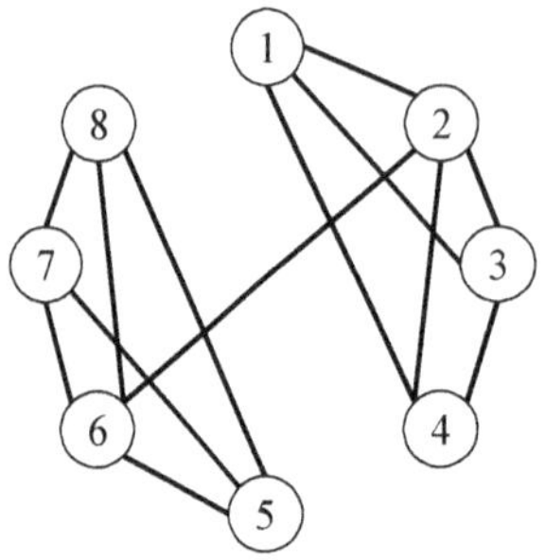

(b) Adjacency matrix

	1	2	3	4	5	6	7	8
1		1	1	1				
2	1		1	1		1		
3	1	1		1				
4	1	1	1					
5								
6		1						
7								
8								

Figure 5.4: A toy example of (a) graph and (b) adjacency matrix. These two show the same topology.

edges. This idea of making labels consistent with edge connectivity is called *graph smoothness.*

Graph Smoothness: Formulation

Let $\boldsymbol{z}$ be a binary vector with the size of N, and we now think to assign each element of $\boldsymbol{z}$ to one of N nodes in a graph as a label. That is, each element in $\boldsymbol{z}$ takes +1 or -1, i.e. $\boldsymbol{z} \in \{+1, -1\}^N$, and each node has either +1 or -1.

Now let us consider how we can set up these binary values in $\boldsymbol{z}$. In order to do this, we can think about the two cases for two nodes: 1) with an edge or 2) without any edge, as follows:

1. **With an edge:** We need to think that the values (labels) of any two nodes connected by an edge should be similar. This automatically leads to considering the following difference between two nodes which should be small (or simply zero) if the corresponding two nodes are connected by some edge:

$$\mathrm{Diff}(\boldsymbol{z}_i, \boldsymbol{z}_j), \tag{5.50}$$

where $\mathrm{Diff}(a, b)$ shows a difference (or distance) between two scalar values a and b.

2. **Without an edge:** If two nodes are not connected by any edge, we can think that we have no information on the similarity between the two nodes. Note that no edges do not mean that there are no similarity between the connected two nodes, and instead we can think that we cannot have any information on similarity between two nodes. Then we do not have to think about the similarity between these nodes (or non-edge nodes).

The above two points can be summarized into the following formulation:

$$\min_{\boldsymbol{z}} \sum_{ij} \boldsymbol{W}_{ij} \mathrm{Diff}(\boldsymbol{z}_i, \boldsymbol{z}_j). \tag{5.51}$$

That is, if there is an edge between node i and node j, $\boldsymbol{W}_{ij}$ is one, and $\text{Diff}(\boldsymbol{z}_i, \boldsymbol{z}_j)$ remains in (5.51), while if no edges, $\boldsymbol{W}_{ij}$ is zero, by which (5.51) becomes zero for such node i and node j.

Thus, the concept, graph smoothness (which makes given node labels consistent with edge connectivity of the given graph) can be implemented by (5.51) (see Chapter 8 of [70] for more detailed explanation of graph smoothness).

In (5.51), there are many possibilities for $\text{Diff}(\boldsymbol{z}_i, \boldsymbol{z}_j)$, i.e. the difference between two elements $\boldsymbol{z}_i$ and $\boldsymbol{z}_j$. However generally we can write this distance as L^p norm, using arbitrary p, as follows:

$$\text{Diff}(\boldsymbol{z}_i, \boldsymbol{z}_j) = ||\boldsymbol{z}_i - \boldsymbol{z}_j||^p, \tag{5.52}$$

while practically we use $p = 2$, i.e. the squared distance. Furthermore, for $p = 2$,

$$\sum_{ij} \boldsymbol{W}_{ij}||\boldsymbol{z}_i - \boldsymbol{z}_j||^2 \quad = \quad \boldsymbol{z}^{\mathsf{T}}\boldsymbol{L}\boldsymbol{z}, \tag{5.53}$$

where $\boldsymbol{L}$ is the graph Laplacian of given graph $\boldsymbol{G}$.

Using this concept, we can consider two paradigms of machine learning: supervised and unsupervised learning, for nodes in a given graph.

Supervised learning generally needs to separate given data into training and test (or evaluation) data to train models by the training data and select the most accurate model for the test data. This is because of avoiding overfitting and also keeping generalization. However, in a graph, we are unable to divide nodes into training and test data so easily, because nodes are connected by edges. Thus, for nodes in a graph, instead of dividing the given data into training and testing, considering practical situations, the problem setting is that part of nodes are unlabeled. In other words, only part of nodes in a given graph have labels and the rest are unlabeled. This setting is called *semisupervised learning* (or more precisely *semisupervised classification*), and particularly for a graph, this learning is called *label propagation*.

On the other hand, we here raise clustering for unsupervised learning, which would be clustering nodes, based on matrix computation, by which clustering nodes of a graph is called *spectral clustering*.

(Semi)supervised Learning: Label Propagation

We focus on binary labels, and then binary vector $\boldsymbol{z}$ is the parameter to be estimated for label propagation. The input is two: 1) given graph $\boldsymbol{G}$, which is equivalent to adjacency matrix $\boldsymbol{W}$, and 2) label $\boldsymbol{y}$, which is partially given.

For label propagation, we then have to satisfy the following two points at the same time:

1. **Label consistency** Label vector $\boldsymbol{z}$ should be consistent with given true labels, i.e. $\boldsymbol{y}$. This can be written as follows:

$$\min_{\boldsymbol{z}} \text{Diff}(\boldsymbol{z}, \boldsymbol{y}). \tag{5.54}$$

2. **Graph smoothness** Also label vector $\boldsymbol{z}$ should be consistent with edge connectivity of the given graph, and this can be evaluated by graph smoothness. As shown in (5.51), graph smoothness is already defined by using given graph (adjacency matrix) $\boldsymbol{W}$ and label vector $\boldsymbol{z}$. This can be written here again:

$$\min_{\boldsymbol{z}} \sum_{ij} \boldsymbol{W}_{ij} \text{Diff}(\boldsymbol{z}_i, \boldsymbol{z}_j). \tag{5.55}$$

Then as already shown by (5.53), if we use the squared distance for $\text{Diff}(\boldsymbol{z}_i, \boldsymbol{z}_j)$ as follows,

$$\text{Diff}(\boldsymbol{z}_i, \boldsymbol{z}_j) = ||\boldsymbol{z}_i - \boldsymbol{z}_j||^2, \tag{5.56}$$

(5.55) can be simply written as

$$\min_{\boldsymbol{z}} \boldsymbol{z}^\top \boldsymbol{L} \boldsymbol{z}. \tag{5.57}$$

Label propagation can be an optimization problem to satisfy both of the above. That is, one of the above two is the objective function and the other is the constraint, and vice versa. For example, we can set the label consistency as the objective function and the graph smoothness as the constraint. Then, using the squared distance for Diff$(*, *)$ and also using (5.53), a possible way of formulating the problem can be given as follows:

$$\arg\min_{\boldsymbol{z}} (\boldsymbol{z} - \boldsymbol{y})^\top (\boldsymbol{z} - \boldsymbol{y}) \text{ subject to } \boldsymbol{z}^\top \boldsymbol{L} \boldsymbol{z} \leq \text{Constant}, \tag{5.58}$$

By relaxing the binary constraint on $\boldsymbol{z}$ so that elements of $\boldsymbol{z}$ can be real-valued, we can use the method of Lagrange multipliers. The optimization problem is a quadratic problem, which can be analytically solved, resulting in the following estimation rule of $\boldsymbol{z}$:

$$\hat{\boldsymbol{z}} = (\boldsymbol{I} + \lambda \boldsymbol{L})^{-1} \boldsymbol{y}. \tag{5.59}$$

We note that this is just one possible way for label propagation, and we can formulate the label propagation problem in many other ways, by formulating the problem so that both (5.54) and (5.55) should be satisfied.

See Chapter 8 of [70] for various problem formulations of label propagation, and also the derivation of the above analytical solution (5.59).

Node Clustering (Spectral Clustering)

As shown in clustering in Section 5.2.1, clustering instances into several groups is the problem of estimating cluster assignment matrix $\boldsymbol{Z}$, where $\boldsymbol{Z}_{ik}$ is one if instance i is assigned to cluster k; otherwise zero. This idea can be applied to clustering nodes in a graph, where nodes are instances.

Also in Section 5.2.1, the formulation (5.46) of constrained K-means clustering has the objective function (5.39) and the constraint term (given by (5.44)). For a graph, the objective function can be represented by graph smoothness, and the constraint term can be the same as the constrained K-means clustering. We will explain this below:

Procedure 5.4: SPECTRAL CLUSTERING.

Input: Adjacency matrix $\boldsymbol{W}$
Output: Cluster assignment matrix $\boldsymbol{Z}$

```
Compute graph Laplacian L from W.;
Estimate Z by solving the eigenvalue problem, given in (5.65).;
Run ClusteringPostProcessing(Ẑ);
Output finally estimated Ẑ.;
```

1. **Graph smoothness** We can start with binary clustering, i.e. dividing nodes into two groups. Then, as shown in Section 5.2.2, the idea of graph smoothness was to make node labels consistent with the edges of the given graph. In clustering, the labels are clusters or groups. This means that graph smoothness exactly implements clustering nodes. Thus by using the squared distance on the difference in labels between two nodes, we can have the following formulation for graph smoothness:

$$\boldsymbol{z}^\top \boldsymbol{L} \boldsymbol{z}, \tag{5.60}$$

where again $\boldsymbol{L}$ is the graph Laplacian of given graph $\boldsymbol{G}$.

In fact this formulation can be extended to clustering given nodes into multiple groups, by using cluster assignment matrix $\boldsymbol{Z}$, as follows:

$$\mathrm{trace}(\boldsymbol{Z}^\top \boldsymbol{L} \boldsymbol{Z}). \tag{5.61}$$

See Chapter 8 of [70] for derivation of this formulation.

2. **Constrained clustering** In Section 5.2.1, we have showed constrained K-means clustering with the constraint, which is given by (5.43) or (5.44). They are the constraint which sets the cluster size (the number of elements) of each cluster at a certain size. That is, this constraint request the balanced clusters. We can use this constraint for clustering nodes in a graph as well:

$$\boldsymbol{Z}^\top \boldsymbol{Z} = \boldsymbol{D}, \tag{5.62}$$

where $\boldsymbol{D}$ is a diagonal matrix, in which each diagonal element specifies the size of the corresponding cluster.

Thus overall clustering nodes in a graph can be formulated as the following optimization problem:

$$\min_{\boldsymbol{Z}} \quad \mathrm{trace}(\boldsymbol{Z}^\top \boldsymbol{L} \boldsymbol{Z}) \text{ subject to } \boldsymbol{Z}^\top \boldsymbol{Z} = \boldsymbol{D}. \tag{5.63}$$

Readers can easily see that this formulation is very similar to that of constrained K-means clustering, shown in (5.46).

Then again by relaxing the binary constraint of the cluster assignment matrix, we can have the following Lagrangian to solve the optimization problem (5.63).

$$L(\boldsymbol{Z}, \lambda) = \mathrm{trace}(\boldsymbol{Z}^\top \boldsymbol{L} \boldsymbol{Z}) - \lambda\, \mathrm{trace}(\boldsymbol{Z}^\top \boldsymbol{Z} - \boldsymbol{D}). \tag{5.64}$$

Then by setting the partial derivative of this Lagrangian with respect to $\boldsymbol{Z}$ to zero, we can estimate cluster assignment matrix $\boldsymbol{Z}$ just by solving the following eigenvalue problem:

$$\boldsymbol{L}\boldsymbol{Z} = \lambda \boldsymbol{Z}. \tag{5.65}$$

After solving this eigenvalue problem, we can run the same alternate postprocessing procedure over $\boldsymbol{Z}$ which is written as a pseudocode by **Procedure 5.2** in Section 5.2.1. **Procedure 5.6** shows a pseudocode of the entire procedure of spectral clustering.

The above constraint of the constrained K-means clustering is just one way, while there are many possible constraints, which can be also incorporated into the formulation of (5.63). Interested readers should be referred to Chapter 8 of [70].

5.2.3 Learning Multiple Data Types: Data Integrative Learning

We now consider the inputs which have more than two data types. In particular we focus on instances with two different data types: 1) vectors and 2) nodes in a graph, at the same time.

When we use two different types of data, a straightforward approach is to combine the data directly. For example, we can transform vectors into a graph, and can somehow combine two graphs together to run some machine learning method over the combined a graph. However this is an *ad hoc* approach for combining different types of data and so unacceptable. Instead we need combine data through formulating the optimization problem. In fact the key point of using multiple data types is that we combine data at the level of formulation of the optimization problem, instead of combining data directly. That is, in terms of the objective functions and constraints, we integrate the given two different types of data.

Notations follow those already used in this section: let $\boldsymbol{X}$ be instances (vectors) with features, and $\boldsymbol{W}$ be the input adjacency matrix (equivalent to the given graph).

Supervised Learning: Linear Regression

Let $\boldsymbol{y}$ be the given vector of true labels. Also let $\boldsymbol{z}$ be a vector of assigning labels to instances.

First of all, these two vectors should be consistent:

$$\min_{\boldsymbol{z}} ||\boldsymbol{z} - \boldsymbol{y}||^2 \tag{5.66}$$

For vectors, if we use linear regression with parameters $\boldsymbol{\beta}$ as a supervised learning model and also the method of least squares for estimating $\boldsymbol{\beta}$ of linear regression, the objective function is to minimize the difference between $\boldsymbol{z}$ and $\boldsymbol{X}\boldsymbol{\beta}$, as follows:

$$\min_{\boldsymbol{z},\boldsymbol{\beta}} ||\boldsymbol{z} - \boldsymbol{X}\boldsymbol{\beta}||^2, \tag{5.67}$$

where additionally $\boldsymbol{\beta}$ should be regularized like L^p norm given by (5.12), and for example, we can take $p = 2$, i.e. $||\boldsymbol{\beta}||^2 = 1$.

Also for nodes in a graph, the label assignment vector $\boldsymbol{z}$ should be consistent with the edge connectivity of the given graph, meaning that the graph smoothness of $\boldsymbol{z}$ over $\boldsymbol{W}$ must be minimized:

$$\min_{\boldsymbol{z}} \boldsymbol{z}^\top \boldsymbol{L} \boldsymbol{z}. \tag{5.68}$$

Thus overall, the supervised learning problem can be formulated into the following optimization problem:

$$\begin{array}{ll} \arg\min\limits_{\boldsymbol{z},\boldsymbol{\beta}} & ||\boldsymbol{z} - \boldsymbol{y}||^2 \\ \text{subject to} & ||\boldsymbol{z} - \boldsymbol{X}\boldsymbol{\beta}||^2 < \text{Const.}, ||\boldsymbol{\beta}||^2 = \text{Const.}, \boldsymbol{z}^\top \boldsymbol{L}\boldsymbol{z} < \text{Const.} \end{array} \tag{5.69}$$

This formulation has two parameters, $\boldsymbol{\beta}$ and $\boldsymbol{z}$, and we are unable to estimate both parameters at the same time. However this is a *biconvex* problem, in which the problem of estimating one parameter is convex when the other parameter is fixed. That is, if one parameter is fixed, the other parameter is estimated very efficiently. Thus we can obtain a local solution by using the following alternating procedure (for simplicity we have two parameters $\boldsymbol{a}$ and $\boldsymbol{b}$):

1. $\boldsymbol{a}$ is fixed, and $\boldsymbol{b}$ is estimated.

2. $\boldsymbol{b}$ is fixed, and $\boldsymbol{a}$ is estimated.

This is true of this case, because (5.69) becomes a convex problem when another parameter is fixed, since (5.69) is quadratic in terms of each of its parameters, i.e. both $\boldsymbol{\beta}$ and $\boldsymbol{z}$.

Thus by relaxing the binary vector $\boldsymbol{z}$ to a real-valued vector, the following Lagrangian can be generated for 5.69:

$$\begin{aligned} L(\boldsymbol{z}, \boldsymbol{\beta}) &= (\boldsymbol{z} - \boldsymbol{y})^\top (\boldsymbol{z} - \boldsymbol{y}) + \lambda_v (\boldsymbol{z} - \boldsymbol{X}\boldsymbol{\beta})^\top (\boldsymbol{z} - \boldsymbol{X}\boldsymbol{\beta}) \\ &+ \lambda_b \boldsymbol{\beta}^\top \boldsymbol{\beta} + \lambda_g \boldsymbol{z}^\top \boldsymbol{L} \boldsymbol{z}, \end{aligned} \tag{5.70}$$

where λ_v, λ_b and λ_g are hyperparameters, not necessarily estimated from given data directly.

We can then set the partial derivative of the Lagrangian to zero with respect to either of $\boldsymbol{z}$ and $\boldsymbol{\beta}$, and have the update rule of $\boldsymbol{\beta}$ and $\boldsymbol{z}$, as follows:

$$\frac{\partial L(\boldsymbol{z}, \boldsymbol{\beta})}{\partial \boldsymbol{z}} = 0 \quad \Rightarrow \quad \hat{\boldsymbol{z}} = ((1 + \lambda_v)\boldsymbol{I} + \lambda_g \boldsymbol{L})^{-1} (\boldsymbol{y} + \lambda_v \boldsymbol{X}\boldsymbol{\beta}). \tag{5.71}$$

$$\frac{\partial L(\boldsymbol{z}, \boldsymbol{\beta})}{\partial \boldsymbol{\beta}} = 0 \quad \Rightarrow \quad \hat{\boldsymbol{\beta}} = (\boldsymbol{X}^\top \boldsymbol{X} + \lambda_\beta \boldsymbol{I})^{-1} \boldsymbol{X}^\top \boldsymbol{z}, \tag{5.72}$$

where $\lambda_\beta = \frac{\lambda_b}{\lambda_v}$.

Again in the biconvex problem, we can use this update rule alternately, by which we can have estimated $\boldsymbol{\beta}$ and $\boldsymbol{z}$, which are obtained from given data, i.e. $\boldsymbol{X}$ and $\boldsymbol{W}$. **Procedure 5.5** summarizes the above approach into a pseudocode, where again the problem is biconvex, by which $\boldsymbol{\beta}$ and $\boldsymbol{z}$ are alternately estimated.

Procedure 5.5: DATA INTEGRATIVE LINEAR REGRESSION.

Input: Main matrix: $\boldsymbol{X}$; adjacency matrix: $\boldsymbol{W}$
Output: Label assignment vector: $\boldsymbol{z}$; linear regression coefficient: $\boldsymbol{\beta}$

```
Initialize z and β.;
Compute graph Laplacian L from W.;
repeat
    Update label assignment vector z by (5.71), fixing β.;
    Update weight β by (5.72), fixing z.;
until convergence;
Output finally estimated ẑ and β̂.;
```

Unsupervised Learning: Clustering

We have both vectors and nodes in a graph under this setting. If you regard vectors as main data, nodes in a graph or edges can be regarded as constraints, indicating the nodes connected by an edge should be in the same cluster. That is, as such, edge information are called *must-link* constraints over nodes in a graph, eventually vectors. Also this problem setting is called *semisupervised clustering*, because the setting is clustering and the must-links are supervised information (for clustering).

Let $\boldsymbol{Z}$ be cluster assignment matrix, and clustering means to estimate $\boldsymbol{Z}$ from data. We here consider constrained K-means clustering for vectors and spectral clustering for nodes in a graph.

For vectors, we can have the objective function given by (5.39). That is,

$$\min_{\boldsymbol{Z}} \mathrm{trace}(\boldsymbol{Z}^\top \boldsymbol{E} \boldsymbol{Z}), \tag{5.73}$$

where the element of the i-th row and the j-th column of $\boldsymbol{E}$, i.e. $\boldsymbol{E}_{ij}$, is

$$\boldsymbol{E}_{ij} = \frac{1}{2N} ||\boldsymbol{x}_i - \boldsymbol{x}_j||^2. \tag{5.74}$$

On the other hand, for nodes in a graph, we can use the graph smoothness for the objective function:

$$\min_{\boldsymbol{Z}} \mathrm{trace}(\boldsymbol{Z}^\top \boldsymbol{L} \boldsymbol{Z}). \tag{5.75}$$

Finally we can use the cluster size constraint given by (5.62) which was used for both constrained K-means clustering and spectral clustering, as follows:

$$\boldsymbol{Z}^\top \boldsymbol{Z} \ = \ \boldsymbol{D}, \tag{5.76}$$

where $\boldsymbol{D}$ is a diagonal matrix, in which each diagonal element specifies the size of the corresponding cluster.

Thus for the above three terms, the first two terms can be the objective functions and the last term is a constraint, which can be formulated into the following

Procedure 5.6: DATA INTEGRATIVE CLUSTERING.

Input: Matrix data $\boldsymbol{X}$; Adjacency matrix $\boldsymbol{W}$

Output: Cluster assignment matrix $\boldsymbol{Z}$

1 Compute $\boldsymbol{E}$ in (5.74) and graph Laplacian $\boldsymbol{L}$ from $\boldsymbol{X}$ and $\boldsymbol{W}$, respectively.;
2 Estimate $\hat{\boldsymbol{Z}}$ by solving the eigenvalue problem, given in (5.79).;
3 Run `ClusteringPostProcessing(`$\hat{\boldsymbol{Z}}$`)`.;
4 Output finally estimated $\hat{\boldsymbol{Z}}$.;

optimization problem:

$$\min_{\boldsymbol{Z}} \boldsymbol{Z}^\top \boldsymbol{E}\boldsymbol{Z} \quad \text{subject to} \quad \boldsymbol{Z}^\top \boldsymbol{L}\boldsymbol{Z} < \text{Const.}, \quad \boldsymbol{Z}^\top \boldsymbol{Z} = \boldsymbol{D}. \tag{5.77}$$

We can then derive the following Lagrangian:

$$L(\boldsymbol{Z}) = \boldsymbol{Z}^\top \boldsymbol{E}\boldsymbol{Z} + \lambda_G \boldsymbol{Z}^\top \boldsymbol{L}\boldsymbol{Z} - \lambda_C(\boldsymbol{Z}^\top \boldsymbol{Z} - \boldsymbol{D}), \tag{5.78}$$

where λ_G and λ_C are hyperparameters.

We emphasize that two types of data can be integrated through the optimization problem, in terms of the objective functions and constraints.

By relaxing the binary constraint of cluster assignment matrix $\boldsymbol{Z}$ so that elements of $\boldsymbol{Z}$ can be real-valued, we can use the method of Lagrange multipliers to solve (5.77).

Then this optimization problem results into the following eigenvalue problem:

$$(\boldsymbol{E} + \lambda_G \boldsymbol{L})\boldsymbol{Z} = \lambda_C \boldsymbol{Z} \tag{5.79}$$

Solving the above eigenvalue problem and after postprocessing to change real-valued $\boldsymbol{Z}$ to binary cluster assignment matrix $\boldsymbol{Z}$ shown in Section 5.2.1, we can have cluster assignment. We note that this cluster assignment can be obtained from given both vectors and the graph, where these two inputs are balanced through the formulation of the optimization problem of (5.77). **Procedure 5.6** summarizes the above data integrative approach of clustering into a pseudocode.

5.3 Feature Learning

As mentioned in the beginning of this chapter, we describe feature learning, particularly two types of approaches: 1) feature selection and 2) feature generation. We do not explain the background of feature learning in so details here, and interested readers can be referred to Section 3.3 of [70] for more details on the background of feature learning.

Let $\boldsymbol{f}_i$ be the i-th feature.

Procedure 5.7: UNSUPERVISED FEATURE SELECTION.

Input: Matrix data $\boldsymbol{X}$

Output: Matrix data with less features $\boldsymbol{F}$

Run K-means `Clustering(`$\boldsymbol{X}$`)` (Note: clustering is done over features).;

foreach *cluster* k **do**

  Find the closest feature $\boldsymbol{f}_{f_k}$ where f_k is given by (5.80).;

Output $\boldsymbol{F} = \{\boldsymbol{f}_{f_1}, \ldots, \boldsymbol{f}_{f_K}\}$.;

5.3.1 Feature Selection

Unsupervised Learning: Clustering

In unsupervised learning, we can perform feature selection just by clustering features. That is, we can first run clustering over features and then focus on cluster centers, which are definitely the summary of the information given by data. Thus we can consider the following three steps:

1. **Run clustering** We can run some clustering method, such as K-means clustering, over features (note that not over instances), by which we can have a set of clusters. In fact we can extract one feature out of one cluster as a representative, and so the number of clusters results in the number of selected features. Thus we can decide the number of features to be selected by the number of clusters.

2. **Find the closest feature to the cluster center** Let $\boldsymbol{S}_k$ and $\boldsymbol{\mu}_k$ be the k-th cluster, being obtained by clustering, and the center of the k-th cluster, respectively. For each cluster, say the k-th cluster, we obtain the feature, say $\boldsymbol{f}_{f_k}$, which is closest to the cluster center $\boldsymbol{\mu}_k$, as follows:

$$f_k = \arg\min_{i \in \boldsymbol{S}_k} d(\boldsymbol{f}_i, \boldsymbol{\mu}_k), \tag{5.80}$$

   where $d(\boldsymbol{a}, \boldsymbol{b})$ is the distance between two vectors $\boldsymbol{a}$ and $\boldsymbol{b}$. We can use any distance measure for $d(\boldsymbol{a}, \boldsymbol{b})$, such as (1 - inner product between $\boldsymbol{a}$ and $\boldsymbol{b}$).

3. **Output the obtained representative features** We output $\{\boldsymbol{f}_{f_1}, \ldots, \boldsymbol{f}_{f_K}\}$ as the selected features. Again K is the number of clusters and the number of the selected features.

Procedure 5.7 shows a summary pseudocode of the above three-step process.

Supervised Learning: Two Step Approach

Again even if we have labels, we can run some clustering algorithm over given data to have cluster centers as representative features of clusters and eventually selected features. However in supervised learning, we have label information, which can be used to improve the performance of selecting features.

Then the fundamental idea of selecting features in supervised learning is as follows: supervised learning is to estimate function F between output $\boldsymbol{y}$ and input $\boldsymbol{X}$ so that $\boldsymbol{y} = F(\boldsymbol{X})$. Then a possible assumption is: if we can output $\boldsymbol{y}$ from $\boldsymbol{X}$ without some feature $\boldsymbol{f}_f$, then feature $\boldsymbol{f}_f$ is irrelevant for outputting $\boldsymbol{y}$ from $\boldsymbol{x}$ and so can be removed. This means that we can remove features, which are irrelevant to the process of supervised learning, i.e. outputting $\boldsymbol{y}$ from input $\boldsymbol{X}$.

We can write this more formally. We can first define *Markov blanket* as follows.

Definition 5.1 (Markov Blanket). *Markov blanket $\mathcal{M}_k$ of feature $\boldsymbol{f}_k$ is a set of features which are correlated with feature $\boldsymbol{f}_k$.*

We can then use the Markov blanket to remove features irrelevant to supervised learning as follows:

Proposition 5.1. *If $\boldsymbol{f}_k$ and its Markov blanket $\mathcal{M}_k$ satisfy* (5.81), *we can remove feature $\boldsymbol{f}_k$, since this feature is irrelevant to the relation between $\boldsymbol{y}$ and $\boldsymbol{X}$.*

$$p(\boldsymbol{y}|\mathcal{M}_k, \boldsymbol{f}_k) \quad = \quad p(\boldsymbol{y}|\mathcal{M}_k). \tag{5.81}$$

The (5.81) examines if the probability of labels given Markov blanket of feature $\boldsymbol{f}_k$ can be changed from the probability of labels given not only Markov blanket of feature $\boldsymbol{f}_k$ but also feature $\boldsymbol{f}_k$ itself. Then if they are not changed (or the same), we can say that feature $\boldsymbol{f}_k$ has nothing to do with the probability of labels given Markov blanket of feature $\boldsymbol{f}_k$, by which we can remove $\boldsymbol{f}_k$. That is, for some feature $\boldsymbol{f}_k$, we can generate Markov blanket $\mathcal{M}_k$ and check if this feature is redundant and can remove this feature by using (5.81).

However this is a time-consuming step, because generating Markov blanket $\mathcal{M}_k$ of a feature needs consider the combination of the label and features in each Markov blanket, and also we need compute the same computation to check the irrelevance of each feature by using (5.81).

Thus for a large number of features, we cannot take this method. Instead we need use a simpler approach, which just focuses on the relevance between each feature and the label vector. We can use any measure for checking the relevance, such as the inner product (correlation).

Overall, we can take a two step approach: we first check the relevance of each feature to labels and reduce the given features to a moderate-sized set of features. We can then check if each of the remaining features is irrelevant or not to the correlation between output $\boldsymbol{y}$ and input $\boldsymbol{X}$, more specifically Markov blanket $\mathcal{M}_k$, by using (5.81). That is, we can remove features, which satisfy (5.81). By this process, we can reduce the moderate-sized set to a very limited number of features. Thus overall the feature selection method can be summarized into the following two steps:

1. **Select features relevant to labels** We can first check the relevance between each feature and the label vector, and select the top most relevant features so that we can keep a good size of features to be selected further.

2. **Remove features irrelevant to the relation between $\boldsymbol{X}$ and $\boldsymbol{y}$** We then check the Markov blanket of each of the remaining features, and check

Procedure 5.8: SUPERVISED FEATURE SELECTION.

Input: Matrix data $\boldsymbol{X}$, #features after Step 1 M_1, final #features M_f
Output: Matrix data with M_f features

Select $\boldsymbol{M}_1$ features which are most relevant to the label vector.;
repeat
  Remove feature $\boldsymbol{f}_k$ which is irrelevant to the relation between $\boldsymbol{X}$ and $\boldsymbol{y}$ and so satisfies (5.81).;
until *#features reduces to* M_f;
Output finally obtained M_f features as a matrix.;

if each instance is redundant or irrelevant in terms of the relation between $\boldsymbol{y}$ and $\boldsymbol{X}$.

These two steps are summarized into a pseudocode in **Procedure 5.8**, in which given data matrix $\boldsymbol{X}$ has M features, which are reduced to M_1 features in Step 1 and then further to M_f features in Step 2 finally.

5.3.2 Feature Generation

In feature selection, we did not change features and just select features which would be useful for summarizing the given data (unsupervised learning) or for understanding the relation between labels $\boldsymbol{y}$ and features $\boldsymbol{X}$ (supervised learning). In this section, instead we show an approach of generating new features, which summarize the given data well, by using the given features. We call this approach *feature generation*, and also this approach has been called in several ways, including *dimensionality reduction*. We show *principal component analysis (PCA)* as a representative feature generation method, while note that there are already numerous methods for the same problem.

Principal Component Analysis (PCA)

The description of PCA here is rather brief, and interest readers can check more detail on PCA in Page 94 of [70].

The idea of PCA is to project given data onto an one-dimensional (1D) space so that the variance on the 1D space should be maximized. Fig. 5.5 shows a schematic diagram of this process. Again, PCA maximizes the broadness (variance) of the distribution of the instances on the 1D space, to which the original instances are projected.

We can first assume that the projection onto the one-dimensional space can be done by a linear combination of given data (So the point of PCA is the new feature is represented by a linear combination of the original features). That is, let w_i be the weight over input feature vector $\boldsymbol{f}_i$. The linear combination over features can be written as follows:

$$w_1\boldsymbol{f}_1 + w_2\boldsymbol{f}_2 + \cdots + w_M\boldsymbol{f}_M = \boldsymbol{X}\boldsymbol{w}, \tag{5.82}$$

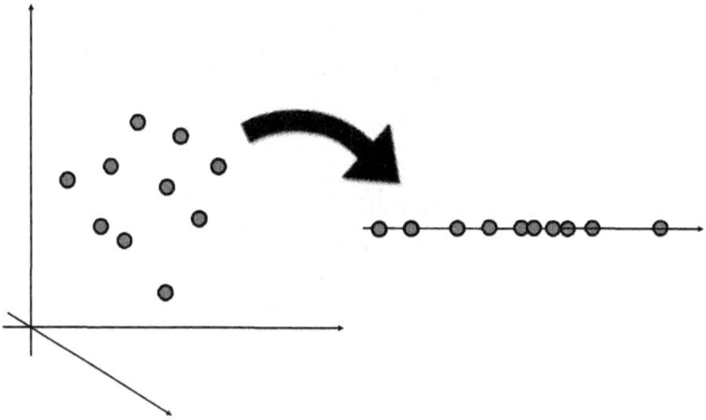

Figure 5.5: Concept of principal component analysis (PCA). Given $\boldsymbol{X}$ is projected on an one-dimensional space, and then the variance of the projected instances is maximized. This picture is taken from [70].

where $\boldsymbol{w} = (w_1, \ldots, w_M)^\top$ and this is the parameter (linear coefficient) to be estimated from data.

Suppose that $\boldsymbol{Xw}$ is already normalized so that the mean is zero, we can compute the variance as follows:

$$(\boldsymbol{Xw})^\top \boldsymbol{Xw} = \boldsymbol{w}^\top \boldsymbol{X}^\top \boldsymbol{Xw}. \tag{5.83}$$

The objective of PCA is to maximize the variance[7], given by the above, i.e. $\boldsymbol{w}^\top \boldsymbol{X}^\top \boldsymbol{Xw}$.

We can then place the constraint over $\boldsymbol{w}$, which should be regularized on the values taken by $\boldsymbol{w}$. For example, the squared sum of $\boldsymbol{w}$ should be a constant:

$$\boldsymbol{w}^\top \boldsymbol{w} = \text{Constant}. \tag{5.84}$$

Then PCA can be formulated as an optimization problem, where the objective function can be written as follows:

$$\max_{\boldsymbol{w}} \boldsymbol{w}^\top \boldsymbol{X}^\top \boldsymbol{Xw} \quad \text{subject to} \quad \boldsymbol{w}^\top \boldsymbol{w} = \text{Constant}. \tag{5.85}$$

Then the Lagrangian $L(\boldsymbol{w})$ of this problem can be written as follows:

$$L(\boldsymbol{w}) = \boldsymbol{w}^\top \boldsymbol{X}^\top \boldsymbol{Xw} - \lambda(\boldsymbol{w}^\top \boldsymbol{w} - \text{Constant}), \tag{5.86}$$

where λ is a Lagrange multiplier.

We set the gradient of Lagrangian $L(\boldsymbol{w})$ with respect to $\boldsymbol{w}$ equal to zero, and this results in an eigenvalue problem, as follows:

$$\frac{dL(\boldsymbol{w})}{d\boldsymbol{w}} = 0 \qquad \Rightarrow \qquad \boldsymbol{X}^\top \boldsymbol{Xw} = \lambda \boldsymbol{w}. \tag{5.87}$$

[7] Note that $\boldsymbol{X}^\top \boldsymbol{X}$ is the matrix to compute the variance for instances in $\boldsymbol{X}$.

Procedure 5.9: PRINCIPAL COMPONENT ANALYSIS.

Input: Matrix data $\boldsymbol{X}$, #features generated: K
Output: new K features

Compute $\boldsymbol{X}^{\top}\boldsymbol{X}$.;
Estimate eigenvectors by solving the eigenvalue problem, given in (6.92).;
Select top K eigenvectors: $\hat{\boldsymbol{w}}_1, \ldots, \hat{\boldsymbol{w}}_K$.;
Project $\boldsymbol{X}$ by the K eigenvectors: $\boldsymbol{X}\hat{\boldsymbol{w}}_1, \ldots, \boldsymbol{X}\hat{\boldsymbol{w}}_K$.;

If we generate K new features, after solving the above eigenvalue problem, we keep the first K eigenvectors[8], i.e. $\boldsymbol{w}_1, \ldots, \boldsymbol{w}_K$, and then new features are obtained by their projection, i.e. $\boldsymbol{X}\boldsymbol{w}_1, \ldots, \boldsymbol{X}\boldsymbol{w}_K$. As such we can generate K new features from the original features in given data $\boldsymbol{X}$. **Procedure 5.9** summarizes, as a pseudocode, this procedure of generating K new features by eigenvalue decomposition and also projection through the eigenvectors.

5.4 Kernel Learning/ Kernel Method

We finally introduce kernel learning, which can cover 1) different types of data, i.e. vectors, nodes in a graph and also combination of them, and 2) different machine learning paradigms, i.e. supervised and unsupervised learning and feature learning.

The point of kernel learning is to use kernel functions, which allow learning models to consider both learning from given data and using already known prior knowledge. In marketing, *kernel ridge regression*, i.e. kernelized ridge regression, is used for identifying products which are highly sold in the fashion apprel industry [99].

5.4.1 Kernel Function

Kernel function $K(\boldsymbol{x}_i, \boldsymbol{x}_j)$ shows a similarity between instances, $\boldsymbol{x}_i$ and $\boldsymbol{x}_j$. The simplest example of the kernel function is so-called the linear kernel, corresponding to the original inner product, while the kernel function can be extended from the linear kernel to a more complicated kernel, which usually defines a high-dimensional space. The kernel function can be designed rather freely by considering the background (or prior) knowledge on the similarity in the application[9]. Thus this extension of the kernel function has a possibility of resulting in a good

[8] In PCA and also its eigenvalue problem, the first eigenvalue (and eigenvector) discriminates given instances most, and eigenvalues (and eigenvectors) are in the descending order in terms of their discrimination ability. Thus we can keep the first K eigenvectors, obtained by solving the eigenvalue problem, since the original objective of PCA was to make the distribution on the project 1D space most diverse.

[9] Note that a kernel function cannot be totally freely designed. That is, a kernel function has to first satisfy the properties of *inner product space*, such as 1. symmetry, 2. linearity and 3. positive semidefiniteness [94, 70]. Thus we can design a kernel function which incorporates background knowledge under those constraints.

Procedure 5.10: DESIGNING KERNEL METHOD.

Input: Original Data $\boldsymbol{X}$, Kernel function: $K(\cdot,\cdot)$
Output: Kernel Method

1 Generate formulation of the problem which has only inner product $\boldsymbol{x}_i^\top \boldsymbol{x}_j$ of two instances $\boldsymbol{x}_i$ and $\boldsymbol{x}_j$ for input $\boldsymbol{X}$.;
2 Replace inner product $\boldsymbol{x}_i^\top \boldsymbol{x}_j$ with kernel function $K(\boldsymbol{x}_i, \boldsymbol{x}_j)$.;

combination between given data and background knowledge. This then might allow to bring a lot of merits for machine learning methods, such as achieving a higher predictive performance than just using given data only or prior knowledge only.

5.4.2 Kernel Learning/ Kernel Method

Again the basic motivation of *kernel methods* in machine learning is to use both given data and prior knowledge (as kernel function). Kernel methods always share the same procedure, in which the basic learning algorithm has the inner product of the input matrix which can be then replaced with a kernel function. This can be summarized as follows:

Step 1. Formulate a basic algorithm with the inner product of input. A key point in this step is that all inputs appearing in the formulation must be represented by the inner product of the input. Then the next step can be easily applied to change the formulation into that of kernel learning.

Step 2. Replace the inner product of input with a kernel function. We then replace all inner products of input appearing in the formulation with kernel functions. By doing this, we can use a more high-dimensional data space than the inner product space by the original input.

Again the above two steps are the common procedure shared by all kernel methods.

5.4.3 Kernel Method Example: Support Vector Machine

In Section 5.2.1, we explained the margin maximization-based algorithm for estimating parameters of linear regression. This linear regression can be extended to a kernel method, called *support vector machine* (SVM), and in fact the idea of SVM is based on the margin maximization. Thus here we will explore the Lagrangian, given in (5.29), further more here. That is,

$$\frac{\partial L(\boldsymbol{\beta}, b, \boldsymbol{\alpha})}{\partial \boldsymbol{\beta}} = 0 \quad \Rightarrow \quad \boldsymbol{\beta} = \sum_i \alpha_i y_i \boldsymbol{x}_i. \tag{5.88}$$

$$\frac{\partial L(\boldsymbol{\beta}, b, \boldsymbol{\alpha})}{\partial b} = 0 \quad \Rightarrow \quad \sum_i \alpha_i y_i = 0. \tag{5.89}$$

By using these two, (5.29) can be written as follows:

$$L(\boldsymbol{\alpha}) = \sum_i \alpha_i - \frac{1}{2}\sum_{i,j} y_i y_j \alpha_i \alpha_j \boldsymbol{x}_i^\top \boldsymbol{x}_j. \tag{5.90}$$

This means that the Lagrangian can be the function of $\boldsymbol{\alpha}$ only, since $\boldsymbol{X}$ and $\boldsymbol{y}$ are given in (5.29), and the problem can be solved by estimating $\boldsymbol{\alpha}$.

Also more importantly the Lagrangian has the inner product $\boldsymbol{x}_i^\top \boldsymbol{x}_j$ in the given data space, while this space can be extend to a more high-dimensional space by replacing the original inner product with kernel function $K(\boldsymbol{x}_i, \boldsymbol{x}_j)$:

$$L(\boldsymbol{\alpha}) = \sum_i \alpha_i - \frac{1}{2}\sum_{i,j} y_i y_j \alpha_i \alpha_j K(\boldsymbol{x}_i, \boldsymbol{x}_j). \tag{5.91}$$

We can then again see that the above procedure also follows the common procedure of kernel methods. That is, SVM can be generated as follows:

Step 1. (5.29) is the original Lagrangian of linear regression based on margin maximization, and this Lagrangian was transformed into (5.90), which has the inner product of the given matrix.

Step 2. Then by replacing the inner product of (5.90) with some kernel function $K(\boldsymbol{x}_i, \boldsymbol{x}_j)$, we can have the formulation with kernel function, as shown in (5.91), which becomes a kernel method, support vector machine (SVM).

SVM is a standard classification method, being well used in many applications. In marketing, for example, SVM is used for modeling the number of sales of smartphone devices by eBay sellers [95].

We showed SVM as an example of kernel learning, while a lot of learning methods can be *kernelized*. For example, ridge regression, K-means clustering and principal component analysis (PCA) can be all transformed into *kernel ridge regression*, *kernel K-means clustering* and kernel PCA, respectively (see Chapter 3 of [70] on the derivation of their kernelization). Note that these three methods are on different machine learning paradigms, i.e. supervised, unsupervised and feature learning, while these three methods can be all regarded as kernel learning. Thus, we can say that kernel learning or kernelization can be applied to any paradigm of machine learning, just by following the above two step procedure.

Also one important advantage of SVM is, once after we estimate $\boldsymbol{\alpha}$, given new instance $\boldsymbol{x}_{\text{new}}$, we can estimate the label of $\boldsymbol{x}_{\text{new}}$, i.e. $\hat{y}_{\text{new}}$, very easily by using (5.88), as follows:

$$\hat{y}_{\text{new}} = \boldsymbol{x}_{\text{new}}^\top \boldsymbol{\beta} = \sum_i \alpha_i y_i \boldsymbol{x}_{\text{new}}^\top \boldsymbol{x}_i. \tag{5.92}$$

Thus, by replacing the inner product with kernel function K, the label can be predicted easily by SVM as follows:

$$\hat{y}_{\text{new}} = \sum_i \alpha_i y_i K(\boldsymbol{x}_{\text{new}}, \boldsymbol{x}_i). \tag{5.93}$$

Procedure 5.11: SUPPORT VECTOR MACHINE.

Input: kernel function:$\boldsymbol{K}$; label:$\boldsymbol{y}$

Output: Coefficients:$\boldsymbol{\alpha}$

1 Estimate $\hat{\alpha}$ by solving (5.15) from $\boldsymbol{X}$, $\boldsymbol{y}$ and λ.;

2 Output finally estimated $\hat{\boldsymbol{\alpha}}$.;

5.4.4 Pros and Cons of Kernel Learning

We here summarize the advantages and disadvantages of kernel learning.

P1. High-dimensional space. A clear merit is that we can design a complex kernel function, which means we can use a higher dimensional space than the original space in which only the inner product of inputs is used. Then high dimensional space would be advantageous, for example, when we discriminate instances in supervised learning, instances are likely to be separated more easily in a higher dimensional space.

P2. Prior knowledge. Again the kernel function can be designed, considering the objective of the application, resulting in incorporating the knowledge of the expert who designs the kernel function. Thus kernel learning is of course run by using given data, while at the same time, the background knowledge of the application can be also incorporated into the machine learning process.

Parameters of kernel learning are estimated by not only given data but also prior knowledge (through designing the kernels to be used), which can be regarded as being equivalent to the objective function (to be optimized) with one or more regularizers. That is, the prior knowledge (kernels) correspond to the regularizers in optimization problems. Thus we can say that kernel functions or prior knowledge play an important role as regularization, to estimate parameters as expected and other merits in optimization, like avoiding overfitting given data.

P3. Sparse model As shown in (5.93), in SVM (kernel learning for regression), we can predict the label of a given unknown instance just by using the label and also the kernel function of each existing (training) instance. This computation may look a heavy burden, since the given unknown instance must be compared with each of all existing training instances. However, the trained $\boldsymbol{\alpha}$ is a sparse vector, meaning that most elements are zero except a small number of non-zero elements. This is because SVM focuses on only support vectors, i.e. instances distributed around the boundary between the given two classes, such as positives and negatives. Then, elements of $\boldsymbol{\alpha}$ corresponding to support vectors are non-zero, while other elements are zero. Thus, the computation of (5.93) can be fast, by using this sparseness property of $\boldsymbol{\alpha}$.

C1. Scalability. A kernel function shows the similarity between two instances, meaning that the function can be a matrix of instances $\times$ instances. This means that if we have a huge size of instances, say one million, it is already totally hard to generate a kernel function over all pairs of instances. Thus kernel learning is not necessarily good for big data but useful for a middle-sized or slightly smaller dataset.

5.5 Key Points

We now raise two key points which can be commonly shared by the (rather simple) machine learning methods we have seen in this chapter.

5.5.1 Squared Distance and L^2 Norm

The squared distance (error) is well used in machine learning. In fact, we can raise examples of the squared error very easily in the explanation even we have made so far:

1. The method of least squares for linear regression, given in (5.9).
2. The distance used in (constrained) K-means clustering, given in (5.37) which is equivalent to computing the variance, given in (5.40).
3. The vector for graph smoothness, given in (5.53), which leads to graph Laplacian.
4. The squared error between the true label and the label to be estimated, in label propagation, given in (5.58).
5. The linear regression (supervised learning) in data integrative learning, given in (5.67).
6. Clustering (unsupervised learning) in data integrative learning, given in (5.74).

There are at least two reasons why the squared distance is well used:

Normal distribution The main reason is from the *normal distribution* assumption over data (on signals, noise, etc.). Some more details on the case of clustering are explained in the description on constrained K-means clustering already.

Derivation in optimization The squared distance in the objective function leads to the convex nature of the function. That is, the squared distance is well used because of the easiness in solving the optimization problem.

In fact the second reason can be shared with the reason why L^2 norm or similar squared term are well used in machine learning. Also for this, we can see the following examples, in the description we have made:

1. Regularization terms of ridge regression, given in (5.14),

Thus, we will use the squared distance and L^2 norm further in the later chapters when we will develop machine learning models and algorithms to solve the problems in marketing.

5.5.2 Rayleigh Quotient

The following term is in general called *Rayleigh quotient*:

$$R(\boldsymbol{M}, \boldsymbol{z}) = \frac{\boldsymbol{z}^\top \boldsymbol{M} \boldsymbol{z}}{\boldsymbol{z}^\top \boldsymbol{z}}, \tag{5.94}$$

where $\boldsymbol{M}$ is (usually symmetric) *positive (semi)definite matrix* and $\boldsymbol{z}$ is a vector. Note that any positive semidefinite matrix is invertible (nonsingular).

A lot of optimization problems in machine learning are equivalent to optimizing the Rayleigh quotient. For example, given $\boldsymbol{M}$, minimizing (5.94), i.e. Rayleigh quotient, is equivalent to the following minimization problem:

$$\min_{\boldsymbol{z}} \boldsymbol{z}^\top \boldsymbol{M} \boldsymbol{z} \quad \text{subject to} \quad \boldsymbol{z}^\top \boldsymbol{z} = \text{Constant}. \tag{5.95}$$

On the other hand, the formulation of the optimization problem of constrained K-means clustering, i.e. (5.46), can be rewritten below:

$$\min_{\boldsymbol{Z}} \quad \text{trace}(\boldsymbol{Z}^\top \boldsymbol{E} \boldsymbol{Z}) \text{ subject to } \boldsymbol{Z}^\top \boldsymbol{Z} = \boldsymbol{D}. \tag{5.96}$$

Obviously this is the same formulation as (5.95). Similarly for example, the formulation of spectral clusering, i.e. (5.63), also has an equivalent formulation of (5.95). In fact this can be true of other machine learning approaches, such as clustering for data integrative learning (5.77), PCA (5.85), etc.

Then all these optimization problems eventually turn into a (generalized) eigenvalue problem. Thus we can say that a lot of machine learning problems can be formulated as a common optimization problem, which is equivalent to optimizing Rayleigh quotient and also solving the generalized eigenvalue problem.

Chapter 6

Machine Learning for Target Marketing

As described in Chapter 3, target marketing has three consecutive steps: segmentation, targeting and positioning (STP) [30, 112]. In this chapter, we explore if each of these three steps can be automatically performed by using machine learning. Techniques of machine learning which will be described here are all based on those shown in Chapter 5, and are kept to be simple and intuitive. According to the three main steps of target marketing, these techniques can be separated into three parts: 1) approaches for customer segmentation which are categorized into two major machine learning paradigms, i.e. unsupervised (clustering) and supervised (classification) learning. 2) methods for ranking the obtained segments purely from viewpoints of machine learning, 3) methods for assisting market positioning, particular focus being placed on conducting the SWOT analysis and generating a perceptual map. Finally we show an approach to find a optimal place for your product in your market, thinking about competitors and also customers.

Below suppose that we are a *company* who wants to perform target marketing for selling our *product* to *customers*. Another important assumption is that we have good enough data on products and customers.

6.1 Data

We first explain the data to be used for machine learning for target marketing. The used data are all matrices, which can be classified by their instances, which are customers or products. That is, one matrix has rows for customers or products and columns for their features. An interesting point is that these two types of data can share the same data spaces, i.e. features, which makes one type of integrative analysis over the data possible. We will explain this point below:

Table 6.1: Example of data matrix $\boldsymbol{X}$ by customers (rows) $\times$ their features (columns). There are at least four types of customer features, and each type has a certain number of features, such as g1, g2, ... for geographic features. These features are for columns, and customers are rows. Thus each elements shows the value of the corresponding feature of each customer.

	geographic			demographic			psychographic			behavioral			...
	g1	g2	...	d1	d2	...	p1	p2	...	b1	b2	...	...
Customer1													
Customer2													
Customer3													
Customer4													
...													

6.1.1 Customers

Traditionally we can use two types of customer matrices, in which rows are both customers, while columns are general (for example demographic) features or survey data. Recently we have other types of data (columns), such as user-generated social tags [76],

Features

The features of this matrix show customer profiles, which can be traditionally generated from four different viewpoint, as explained in Section 3.1. Four different viewpoints are: geographic, demographic, psychographic and behavioral features. Table 6.1 shows a schematic example of a matrix with these four different types of features. Note that currently such user profiles can be generated through online user activities [104]. This matrix is used for generating customer segments, which correspond to clusters of customers.

Also as we mentioned as **Step 1** of SWOT analysis in Section 3.3, we may first run some **feature learning** algorithm over this matrix to keep only the features which should be useful for the remaining analysis, e.g. [100].

Survey

Table 6.2 shows a schematic way of generating this type of matrix. The top table shows that each row is a questionnaire on a product which is answered by the corresponding customer, meaning that each element is an answer of a particular question on a product by a customer. Then for each customer, the answers for all products are summed (averaged) over all products, results in the bottom table. Thus the original table has both information on customers and products, by which the original table allows to connect customers and products. This point will be explained further below.

Table 6.2: (top) Original survey data. (bottom) (Customers × Features)-table obtained by, for each customer, averaging feature values over products.

		Feature 1	Feature 2	Feature 3	...
Customer1	Product A				
	Product B				
	...				
Customer2	Product A				
	Product B				
	...				
Customer3	Product A				
	Product B				
	...				

⇓

	Feature 1	Feature 2	Feature 3	...
Customer1				
Customer2				
...				

Table 6.3: Example matrix of products (rows) × their features (columns).

	durability	annual sales	annual profit	...
Product A				
Product B				
Product C				
Product D				
...				

6.1.2 Products

Also the other type of matrices has rows for products. Similar to customers, this matrix has two types of features as explained below:

Specification

This table has columns for features, which are on product specification, which are objective properties of products, being different from subjective information like answers of questionnaires. Table 6.3 shows a schematic example of this matrix. In fact the features of this table can be obtained only by the specification or statistics of products.

Table 6.4: (top) Original survey data. (middle) Records are sorted, according to products, instead of customers. (bottom) For each product, feature values are summed up over all customers, resulting in a matrix of products × features.

		Feature 1	Feature 2	Feature 3	...
Customer1	Product A				
	Product B				
	...				
Customer2	Product A				
	Product B				
	...				
Customer3	Product A				
	Product B				
	...				

⇓

		Feature 1	Feature 2	Feature 3	...
Product A	Customer1				
	Customer2				
	...				
Product B	Customer1				
	Customer2				
	...				
Product C	Customer1				
	Customer2				
	...				

⇓

	Feature 1	Feature 2	Feature 3	...
Product A				
Product B				
Product C				
...				

Table 6.5: A schematic example of (products × features)-matrix generated from survey (questionnaire) results. This is on digital camera.

	memory size	picture quality	robustness	size	weight
A	10	9	3	2	1
B	4	7	8	5	6
C	3	4	1	9	8
D	8	8	5	1	5
...					

Survey

This table is generated from a survey record matrix, which is the same as the top of Table 6.2. We show a process of generating this matrix from the original survey record matrix. Table 6.4 shows a flow of generating a matrix of products × features from the original survey data. The top of Table 6.4 is the original survey data which is the same as that shown in the top of Table 6.2. We then sort the record (instance) so that customer answers are grouped, according to products. This step is schematically shown in the middle of Table 6.4, where customers are gathered together for each product. Then we sum up (average over) the customer answers for each product. The bottom of Table 6.4 shows the resultant table of products × features. Table 6.5 shows a more realistic example of the final result of Table 6.4.

6.2 Market Segmentation

Market segmentation is to divide customers into a certain number of groups, by using customer data. We use the feature data of customers for generating segmentation. The features of this matrix were mainly divided into four types, as mentioned in Section 3.1: 1) geographic (place the customer is living), 2) demographic (customer life style), 3) psychographic (personalities and lifestyle) and 4) behavioral (product knowledge and buying trend) data. All four types of information can be features of customers (instances). Thus entirely given data can be a matrix $\boldsymbol{X}$, having customers for rows and the above features for columns, meaning that each instance is a vector. Let $\boldsymbol{x}_i$ be the i-th row of $\boldsymbol{X}$, i.e. one instance or customer. Table 6.1 shows a schematic example of given data $\boldsymbol{X}$.

We show methods to automatically perform market segmentation by using machine learning. In fact machine learning (or more generally data science) has been applied to market segmentation, while applied methods are rather simple statistics and machine learning [28].

On the other hand, in this chapter, we show more basic methods, just to show an approach for designing a machine learning model which is matched to given data and problem setting of marketing.

6.2.1 Clustering (Unsupervised Setting)

Market segmentation of customers is, by using $\boldsymbol{X}$, to summarize customers into a limited number of groups, which is exactly equivalent to clustering, the most typical unsupervised machine learning technique. This is rather obvious in the literature of marketing, and already various clustering methods (and cluster analysis) have been widely applied to market segmentation (e.g. classic: [86], reviews: [27, 105]).

We focus on disjoint clustering, which estimates cluster assignment matrix $\boldsymbol{Z}$, as exactly explained in Section 5.2.1, which has customers for rows and clusters for columns: $\boldsymbol{Z}$ is a binary matrix, in which for each row only one column has one; otherwise zero, meaning that the column corresponding one is the cluster to which the row (customer) is assigned.

We clarify the input and output of this problem:

Input:

1. (customers $\times$ features)-matrix $\boldsymbol{X}$. We have N customers, $\boldsymbol{x}_1, \ldots, \boldsymbol{x}_N$, i.e. $\boldsymbol{X} = (\boldsymbol{x}_1, \ldots, \boldsymbol{x}_N)^\mathsf{T}$. Let M be the size of features.

Output:

1. $\boldsymbol{Z}$, segment assignment matrix, defined in Section 5.2.1, which has customers for rows, segments for columns and binary elements, meaning that the value is one if a customer is in the corresponding segment; otherwise zero.

We start with applying a standard clustering method to segmenting customers, by using $\boldsymbol{X}$ only for the input, and then explore other more realistic settings of clustering, in which not only $\boldsymbol{X}$ but also other information are available.

Direct Application of Constrained K-means Clustering

As a starting point, a straightforward approach of running clustering over $\boldsymbol{X}$, for example, (constrained) K-means clustering, which is explained in Section 5.2.1, can be applied to this problem.

We briefly review constrained K-means clustering: the objective is to minimize the distance between each instance, say the i-th instance $\boldsymbol{X}_i$, and the center $\boldsymbol{\mu}_k$ of cluster k, to which the instance is assigned. This is shown in (5.34), and again as follows:

$$\min_{\boldsymbol{Z},\boldsymbol{\mu}} \sum_{k=1}^{K} \sum_{i=1}^{N} \boldsymbol{Z}_{ik} \mathrm{Dist}(\boldsymbol{x}_i, \boldsymbol{\mu}_k). \tag{6.1}$$

If we use the squared distance as shown in (5.36), (6.1) can be written as in (5.37):

$$\min_{\boldsymbol{Z},\boldsymbol{\mu}} \sum_{k=1}^{K} \sum_{i=1}^{N} \boldsymbol{Z}_{ik} ||\boldsymbol{x}_i - \boldsymbol{\mu}_k||^2. \tag{6.2}$$

Then according to (5.38) and (5.39), this can be transformed into the following trace:

$$\sum_{k=1}^{K}\sum_{i=1}^{N} Z_{ik}||\boldsymbol{x}_i - \boldsymbol{\mu}_k||^2 = \text{trace}(\boldsymbol{Z}^\top \boldsymbol{E}\boldsymbol{Z}), \tag{6.3}$$

where as shown in (5.40), the element of the i-th row and the j-th column of $\boldsymbol{E}$, i.e. $\boldsymbol{E}_{ij}$, is the squared distance between two instances $\boldsymbol{x}_i$ and $\boldsymbol{x}_j$, as follows:

$$\boldsymbol{E}_{ij} = \frac{1}{2N}||\boldsymbol{x}_i - \boldsymbol{x}_j||^2. \tag{6.4}$$

On the other hand, the cluster assignment matrix $\boldsymbol{Z}$ must be constrained, because each customer belongs to only one group, meaning that each row has one for only one column and otherwise zero. This can be written as in (5.42). From this observation we can place the constraint over $\boldsymbol{Z}^\top\boldsymbol{Z}$, as in (5.44):

$$\boldsymbol{Z}^\top\boldsymbol{Z} = \boldsymbol{D}, \tag{6.5}$$

where $\boldsymbol{D}$ is the diagonal matrix with each diagonal element specifying the size (number of instances) of the corresponding cluster.

Overall the optimization problem of constrained K-means clustering can be formulated as follows:

$$\min_{\boldsymbol{Z}} \quad \text{trace}(\boldsymbol{Z}^\top\boldsymbol{E}\boldsymbol{Z}) \text{ subject to } \boldsymbol{Z}^\top\boldsymbol{Z} = \boldsymbol{D}. \tag{6.6}$$

By relaxing the binary constraint of $\boldsymbol{Z}$ to have real values in $\boldsymbol{Z}$, the Lagrangian of this optimization problem is

$$L(\boldsymbol{Z}, \lambda) = \text{trace}(\boldsymbol{Z}^\top\boldsymbol{E}\boldsymbol{Z}) - \lambda \text{ trace}(\boldsymbol{Z}^\top\boldsymbol{Z} - \boldsymbol{D}). \tag{6.7}$$

By setting the partial derivative of the Lagrangian with respect to $\boldsymbol{Z}$ to zero, we can have the eigenvalue problem:

$$\boldsymbol{E}\boldsymbol{Z} = \lambda\boldsymbol{Z}. \tag{6.8}$$

After solving this eigenvalue problem, we run the postprocessing algorithm over $\boldsymbol{Z}$ to have discrete values in $\boldsymbol{Z}$. This postprocessing is described as a pseudocode of **Procedure 5.2** in Section 5.2.1. Also K-means clustering itself is summarized into a pseudocode in **Procedure 5.3**. Then once again we wrap up the procedure shown above as a pseudocode in **Procedure 6.1**.

Modification of Constraints

Again market segmentation is equivalent to clustering, and a straightforward manner of using machine learning is to apply constrained K-means clustering to given customer data $\boldsymbol{X}$ to generate customer segmentation as $\boldsymbol{Z}$.

One simple modification is to change (6.5). In fact as shown in (5.43), $\boldsymbol{D}$ specified the size of each of K clusters. We can change the cluster size, $|C_k|$ ($k =$

Procedure 6.1: MARKET SEGMENTATION BY K-MEANS CLUSTERING.

Input: (customers × features)-matrix: $\boldsymbol{X}$
Output: Segmentation assignment matrix: $\boldsymbol{Z}$

Compute $\boldsymbol{E}$ from $\boldsymbol{X}$ by using (6.4).;
Estimate $\hat{\boldsymbol{Z}}$ by solving the eigenvalue problem given in (6.8).;
Run `ClusteringPostProcessing(`$\hat{\boldsymbol{Z}}$`)`.;
Output finally estimated $\hat{\boldsymbol{Z}}$.;

$1, \ldots, K$), according to some prior knowledge, while this change has no effect on the final optimization step, i.e. eigenvalue decomposition.

We then change the constraint as follows:

$$\mathrm{trace}(\boldsymbol{Z}^\top \boldsymbol{D} \boldsymbol{Z}) \leq \mathrm{Constant}, \tag{6.9}$$

where $\boldsymbol{D}$ is not a matrix of clusters × clusters, but a matrix of customers × customers. We can design $\boldsymbol{D}$ freely by using prior knowledge. For example, simply if $\boldsymbol{D}$ is a diagonal matrix:

$$\boldsymbol{D} = \begin{pmatrix} d_1 & 0 & \cdots & 0 \\ 0 & d_2 & 0 & \vdots \\ \vdots & 0 & \ddots & 0 \\ 0 & \cdots & 0 & d_N \end{pmatrix}. \tag{6.10}$$

This emphasizes each customer by the corresponding diagonal value, e.g. the importance of each customer. On the other hand, we can use some prior knowledge on customer connection for generating $\boldsymbol{D}$, and then the resultant constraint becomes the number of customer connections.

Then the optimization problem can be formulated as follows:

$$\min_{\boldsymbol{Z}} \quad \mathrm{trace}(\boldsymbol{Z}^\top \boldsymbol{E} \boldsymbol{Z}) \text{ subject to } \mathrm{trace}(\boldsymbol{Z}^\top \boldsymbol{D} \boldsymbol{Z}) \leq \mathrm{Constant}. \tag{6.11}$$

Eventually the final step can be the following generalized eigenvalue problem:

$$\boldsymbol{E}\boldsymbol{Z} = \lambda \boldsymbol{D}\boldsymbol{Z}. \tag{6.12}$$

Thus by solving the generalized eigenvalue problem of (6.12), we can have cluster assignment (segmentation) $\boldsymbol{Z}$.

Grouping Features: Incorporating Group Norm on $\boldsymbol{E}$

Customer data $\boldsymbol{X}$ has four different types of features: geographic, demographic, psychographic and behavioral data. Also we may add more different types of features. Then it might be good to consider the different groups, such as geographic, demographic, etc., over different types of features.

To explain this further, we use the following notations: Let $\boldsymbol{x}_{ij}$ be the j-th feature of instance $\boldsymbol{x}_i$. Let $\boldsymbol{G}^{(g)}$ $(g = 1, \ldots, G)$ be groups of features, and so the given set of features $\boldsymbol{F} = (\boldsymbol{f}_1, \ldots, \boldsymbol{f}_M) = (\boldsymbol{G}^{(1)}, \cdots, \boldsymbol{G}^{(G)})$. We note that $\boldsymbol{G}_g$ is a matrix unless this group has only one feature.

There must be numerous ways of considering the grouping effects, while one way is to change some part of the formulation of K-means clustering, given by (6.6). For example, we can change $\boldsymbol{E}$, given by (6.4), directly by considering the feature groups.

Then originally $\boldsymbol{E}$ was the squared distance (simply we now remove $\frac{1}{2N}$ in (6.4) for simplicity without loss of generality) over all features equally which can be, in general, written by the idea of L^p-norm, instead of just the squared distance, as follows:

$$\begin{aligned} \boldsymbol{E}_{ij} &= ||\boldsymbol{x}_i - \boldsymbol{x}_j||^p && (6.13) \\ &= (\sum_k |\boldsymbol{x}_{ik} - \boldsymbol{x}_{jk}|^p)^{\frac{1}{p}}, && (6.14) \end{aligned}$$

where again $\boldsymbol{x}_{ik}$ is the k-th element of vector $\boldsymbol{x}_i$ (k-th feature of instance $\boldsymbol{x}_i$).

We can then consider the L^p norm for each group and also different norms over the groups. That is, for example, one possible modification is that L^p-norm works only in groups, and another norm, say L^q-norm, can be used between groups:

$$\boldsymbol{E}_{ij} = (\sum_g^G (\sum_{k \in \boldsymbol{G}^{(g)}} |\boldsymbol{x}_{ik} - \boldsymbol{x}_{jk}|^p)^{\frac{q}{p}})^{\frac{1}{q}}, \quad (6.15)$$

where as well as p, q is a hyperparameter to define how groups should be considered when computing $\boldsymbol{E}$, i.e. the distance between instances. Thus p and q can be arbitrary selected. First note that as we can easily see, if we use $p = q = 2$, (6.15) is the same as the original $\boldsymbol{E}$ given by (6.4). We can use different values for p and q to have grouping effects.

It would be hard to optimize the values of p and q automatically, while an empirical manner to decide them is to explore the optimal values by checking the performance on test/evaluation data[1].

Note that $\boldsymbol{E}$ can be computed from given $\boldsymbol{X}$, and so the procedure of constrained K-means algorithm can be kept totally the same. That is, we can first compute $\boldsymbol{E}$ from given $\boldsymbol{X}$, and then the formulation is given by (6.6), and the procedure after this can be the same. **Procedure 6.3** shows a pseudocode of this procedure.

[1]Practically cross-validation is often used for this purpose. Cross-validation is to divide given data into several subdata, called *folds*, and then repeat the following procedure, taking each fold as test data: one fold is test data and the rest folds are training data, and then the performance of cross-validation is obtained by taking the average over all test data. We change the values of p and q to decide these values so that the highest performance of cross-validation should be obtained.

Procedure 6.2: MARKET SEGMENTATION BY K-MEANS CLUSTERING WITH GROUP NORM.

Input: (customers $\times$ features)-matrix $\boldsymbol{X}$; parameters p and q
Output: Segmentation assignment matrix $\boldsymbol{Z}$

Compute $\boldsymbol{E}$ from $\boldsymbol{X}$, p and q, by using (6.15).;
Estimate $\hat{\boldsymbol{Z}}$ by solving the generalized eigenvalue problem given by (6.12).;
Run `ClusteringPostProcessing(`$\hat{\boldsymbol{Z}}$`)`.;
Output finally estimated $\hat{\boldsymbol{Z}}$.;

Grouping Features: Weights over Feature Groups

Another approach to consider the difference between groups is to separate groups totally and place some weight over each group in the objective function in the formulation given by (6.6). First let $\boldsymbol{E}^{(g)}$ be the matrix with the (i, j)-element, $\boldsymbol{E}^{(g)}_{i,j}$, which is the squared distance between the i-th instance and the j-th instance by the features of the g-th group only, as follows:

$$\boldsymbol{E}^{(g)}_{ij} = \sum_{k \in \boldsymbol{G}_g} ||\boldsymbol{x}_{ik} - \boldsymbol{x}_{jk}||^2 \tag{6.16}$$

That is, again $\boldsymbol{E}^{(g)}$ is the difference between two instances by group $\boldsymbol{G}^{(g)}$ only.

Then not using all features but we just focus on the g-th feature group only, i.e. $\boldsymbol{G}^{(g)}$ only, the formulation of (6.6) can be written as follows:

$$\min_{\boldsymbol{Z}} \quad \mathrm{trace}(\boldsymbol{Z}^{\mathsf{T}} \boldsymbol{E}^{(g)} \boldsymbol{Z}) \text{ subject to } \boldsymbol{Z}^{\mathsf{T}} \boldsymbol{Z} = \boldsymbol{D}. \tag{6.17}$$

We have to consider not only $\boldsymbol{G}^{(g)}$ but all feature groups. Then we can place an assumption that feature groups are not equal on generating clusters. In other words, some feature group would be useful, while another feature group might not be so useful for estimating segments. The problem here is to choose useful groups more than useless ones. We then use weights $\boldsymbol{w} = (w_1, \ldots, w_G)^{\mathsf{T}}$ over G groups $\boldsymbol{G}$, where w_g takes a positive value between zero and one, indicating that larger weights are more useful.

1. Setting weights manually

If we have some background knowledge over the feature groups in data, we might decide weights w_g $(g = 1, \ldots, G)$ manually. That is, we can assign weight values so that for example $\sum_g w_g = 1$. Then we can compute $\boldsymbol{E}$ by using $\boldsymbol{E}^{(g)} (g = 1, \ldots, G)$ and w_g $(g = 1, \ldots, G)$, as follows:

$$\boldsymbol{E} = \sum_g w_g \boldsymbol{E}^{(g)}. \tag{6.18}$$

We can use this $\boldsymbol{E}$ with the regular constrained K-means clustering. That is, this $\boldsymbol{E}$ can be used for (5.46) and the subsequent procedure of the constrained

K-means clustering can be run as it is.

2. Learning weights

Another approach is to estimate weights from given data. We here show one possible way for realizing this approach. Note that we show just one possible (maybe simplest) approach for this problem, while there are other solutions in this framework, which will be described in the last of this part. First we need to set some constraint over w_g, for example, L^p-norm:

$$||\boldsymbol{w}||^p = |\sum_g |w_g|^p|^{\frac{1}{p}} = \text{Constant}. \tag{6.19}$$

Furthermore (in reality, in order to easily estimate w_g in our formulation of the optimization problem later), we can take the squared form or L^2 norm, i.e. $p = 2$, and Constant$= 1$. That is,

$$||\boldsymbol{w}||^2 = \sum_g |w_g|^2 = 1. \tag{6.20}$$

The w_g is a weight, meaning that we would like to set a formulation so that weight w_g should be larger if group $\boldsymbol{G}^{(g)}$ is more useful. In fact in (6.18), if group G is more useful, the distance between instances should be properly close and eventually $\boldsymbol{E}^{(g)}$ would be small. This reveals that (6.18) is not well defined, because if group $\boldsymbol{G}_g$ is more useful, $\boldsymbol{E}^{(g)}$ has to be smaller, for which w_g cannot be larger.

Then we need to make the values of $\boldsymbol{E}^{(g)}$ reverse. That is, $\boldsymbol{E}^{(g)}$ was distances (smaller better), and so we define similarities (larger better). In reality we can use, for example (as the simplest case), cosine similarity between two instances, i.e. $\boldsymbol{x}_i$ and $\boldsymbol{x}_j$, for the corresponding element of the similarity matrix, $\boldsymbol{S}^{(g)}_{ij}$, as follows:

$$\boldsymbol{S}^{(g)}_{ij} = \sum_{k|k\in\boldsymbol{G}^{(g)}} \boldsymbol{x}_{ik}\boldsymbol{x}_{jk} = (\boldsymbol{x}^{(g)}_i)^\top \boldsymbol{x}^{(g)}_j, \tag{6.21}$$

where $\boldsymbol{x}^{(g)}_j$ is instance $\boldsymbol{x}_i$ with only features in group $\boldsymbol{G}^{(g)}$.

Thus we can write $\boldsymbol{S}^{(g)}$ by using $\boldsymbol{X}$ as follows:

$$\boldsymbol{S}^{(g)} = \boldsymbol{X}^{(g)}(\boldsymbol{X}^{(g)})^\top \tag{6.22}$$

where $\boldsymbol{X}^{(g)}$ have instances with only features in group $\boldsymbol{G}^{(g)}$. We note that $\boldsymbol{S}^{(g)}$ is a symmetric matrix.

Then as shown in the objective function of (6.17), we can consider the following trace of the quadratic form of $\boldsymbol{Z}$:

$$\text{trace}(\boldsymbol{Z}^\top \boldsymbol{S}^{(g)} \boldsymbol{Z}). \tag{6.23}$$

We can prove that this term has a good property to use the term for our optimization problem. First we can show the following proposition:

Proposition 6.1. *$\boldsymbol{a}^\top \boldsymbol{S}^{(g)} \boldsymbol{a}$ is always non-negative for any arbitrary vector $\boldsymbol{a}$.*

Proof.

$$\begin{aligned} \boldsymbol{a}^\top \boldsymbol{S}^{(g)} \boldsymbol{a} &= \boldsymbol{a}^\top \boldsymbol{X}^{(g)} (\boldsymbol{X}^{(g)})^\top \boldsymbol{a} && (6.24) \\ &= ((\boldsymbol{X}^{(g)})^\top \boldsymbol{a})^\top (\boldsymbol{X}^{(g)})^\top \boldsymbol{a} && (6.25) \\ &= ((\boldsymbol{X}^{(g)})^\top \boldsymbol{a})^2 \geq 0. && (6.26) \end{aligned}$$

□

Thus by using this proposition, we can show the following proposition:

Proposition 6.2. *For arbitrary matrix $\boldsymbol{Z}$*

$$\textit{trace}(\boldsymbol{Z}^\top \boldsymbol{S}^{(g)} \boldsymbol{Z}) \geq 0. \tag{6.27}$$

Proof. Let $\boldsymbol{Z} = (\boldsymbol{a}_1, \ldots, \boldsymbol{a}_K)$, where $\boldsymbol{a}_k$ $(k = 1, \ldots, K)$ are arbitrary vectors and K is also an arbitrary number. The (k, k)-th element of $\boldsymbol{Z}^\top \boldsymbol{S}^{(g)} \boldsymbol{Z}$ is $\boldsymbol{a}_k^\top \boldsymbol{S}^{(g)} \boldsymbol{a}_k$. From **Proposition 6.1**,

$$\boldsymbol{a}_k^\top \boldsymbol{S}^{(g)} \boldsymbol{a}_k \geq 0. \tag{6.28}$$

Thus

$$\text{trace}(\boldsymbol{Z}^\top \boldsymbol{S}^{(g)} \boldsymbol{Z}) \geq 0. \tag{6.29}$$

□

This means, as we expected, trace($\boldsymbol{Z}^\top \boldsymbol{S}^{(g)} \boldsymbol{Z}$) takes always a positive value, and also becomes larger if group $\boldsymbol{G}^{(g)}$ is more useful. Then we can define the term to be *maximized*, as follows:

$$\sum_g w_g \text{trace}(\boldsymbol{Z}^\top \boldsymbol{S}^{(g)} \boldsymbol{Z}). \tag{6.30}$$

In short this means that learning weights over a linear function (which hereafter we call *weight learning*) can be solved by a maximization problem.

Also we need to use the following constraints for parameters to be estimated, i.e. $\boldsymbol{Z}$ and $\boldsymbol{w}$, as follows:

$$\boldsymbol{Z}^\top \boldsymbol{Z} = \boldsymbol{D} \text{ and } ||\boldsymbol{w}||^2 = 1. \tag{6.31}$$

Finally we can formulate our problem of estimating weights over groups, as the following maximization problem:

$$\max_{\boldsymbol{Z},\boldsymbol{w}} \sum_g w_g \text{trace}(\boldsymbol{Z}^\top \boldsymbol{S}^{(g)} \boldsymbol{Z}) \text{ subject to } \boldsymbol{Z}^\top \boldsymbol{Z} = \boldsymbol{D} \text{ and } ||\boldsymbol{w}||^2 = 1, \tag{6.32}$$

where again there are two parameters to be estimated, i.e. cluster assignment matrix $\boldsymbol{Z}$ and weights over groups $\boldsymbol{w}$.

We can then derive the following Lagrangian from this formulation, and now we relax the binary constraint of $\boldsymbol{Z}$ so that $\boldsymbol{Z}$ can be real-valued:

$$L(\boldsymbol{Z}, \boldsymbol{w}) = \sum_g w_g \mathrm{trace}(\boldsymbol{Z}^\top \boldsymbol{S}^{(g)} \boldsymbol{Z}) - \lambda_1 \mathrm{trace}(\boldsymbol{Z}^\top \boldsymbol{Z} - \boldsymbol{D}) - \frac{\lambda_2}{2}(\sum_g (w_g)^2 - 1), \tag{6.33}$$

where λ_1 and λ_2 are hyperparameters.

This Lagrangian has two parameters, $\boldsymbol{Z}$ and $\boldsymbol{w}$, and we are unable to estimate both at the same time. However this is a biconvex problem, which as briefly explained in Section 5.2.3, we can obtain a local optimum solution by using the following alternating procedure:

1. $\boldsymbol{w}$ is fixed, and $\boldsymbol{Z}$ is estimated.

2. $\boldsymbol{Z}$ is fixed, and $\boldsymbol{w}$ is estimated.

Thus in order to consider the situation that one parameter is fixed and the other parameter is estimated, we show the results obtained by that the partial derivative of Lagrangian, given by (6.33), with respect to $\boldsymbol{Z}$ (or $\boldsymbol{w}$) is set at zero, as follows:

$$\frac{\partial L(\boldsymbol{Z}, \boldsymbol{w})}{\partial \boldsymbol{Z}} = 0 \quad \Rightarrow \quad \boldsymbol{S}\boldsymbol{Z} = \lambda_1 \boldsymbol{Z} \tag{6.34}$$

$$\frac{\partial L(\boldsymbol{Z}, \boldsymbol{w})}{\partial \boldsymbol{w}} = 0 \quad \Rightarrow \quad \mathrm{trace}(\boldsymbol{Z}^\top \boldsymbol{S}^{(g)} \boldsymbol{Z}) = \lambda_2 w_g \tag{6.35}$$

Please note that (6.34) results in an eigenvalue problem, where $\boldsymbol{S}$ can be computed as follows:

$$\boldsymbol{S} = \sum_g w_g \boldsymbol{S}^{(g)}. \tag{6.36}$$

Also for the both sides of (6.35), taking the squared sum and using (6.20), we can have the following:

$$\sqrt{\sum_g (\mathrm{trace}(\boldsymbol{Z}^\top \boldsymbol{S}^{(g)} \boldsymbol{Z}))^2} = \lambda_2 \tag{6.37}$$

That is,

$$\hat{w}_g = \frac{\mathrm{trace}(\boldsymbol{Z}^\top \boldsymbol{S}^{(g)} \boldsymbol{Z})}{\sqrt{\sum_g (\mathrm{trace}(\boldsymbol{Z}^\top \boldsymbol{S}^{(g)} \boldsymbol{Z}))^2}}. \tag{6.38}$$

Procedure 6.3 shows a pseudocode of this procedure for estimating $\boldsymbol{Z}$ and $\boldsymbol{w}$ until convergence. As such we can estimate weights over feature groups from given data, at the same time doing segmentation over given data.

Procedure 6.3: SEGMENTATION WITH LEARNING GROUP WEIGHTS.

Input: (customers × features)-matrix $\boldsymbol{X}$
Output: Segmentation assignment matrix $\boldsymbol{Z}$; weights over feature groups $\boldsymbol{w}$

```
Preparation: compute S^(g) (g = 1, ..., G) from X through
    E^(g) (g = 1, ..., G).;
repeat
    /* Estimate Z, fixing w.                                   */
    Compute S by (6.36).;
    Estimate Ẑ by solving the eigenvalue problem given by (6.34).;
    Run ClusteringPostProcessing (Ẑ).;
    /* Estimate w, fixing Z.                                   */
    Estimate ŵ by running (6.38) (or (??)).;
until convergence;
Output finally estimated Ẑ and ŵ.;
```

Another regularizer: In the above, we used $p = 2$ for (6.19), i.e. (6.20). This is equivalent to the regularizer in ridge (linear) regression (ridge-type regularizer). We can examine another regularizer by changing p.

For example, we can take $p = 1$, meaning that a LASSO type regularizer. This setting, $p = 1$, means $\sum_g w_g = 1$, which leads to the sparseness over w_g. Our problem is to maximize

$$\sum_g w_g \mathrm{trace}(\boldsymbol{Z}^\top \boldsymbol{S}^{(g)} \boldsymbol{Z}), \tag{6.39}$$

where $\mathrm{trace}(\boldsymbol{Z}^\top \boldsymbol{S}^{(g)} \boldsymbol{Z})$ is always non-negative. Thus by using a non-negative function of g, f_g, our maximization problem can be written as follows:

$$\sum_g w_g f_g. \tag{6.40}$$

Then under $\sum_g w_g = 1$, the sparse solution is to pick up only one g', which gives the maximum among all possible f_g $(g = 1, \dots, G)$:

$$g' = \arg\max_g f_g \tag{6.41}$$

That is, in our problem, we obtain the solution of w_g as follows:

$$\hat{w}_{g'} = 1 \qquad \text{if} \quad g' = \arg\max_g \mathrm{trace}(\boldsymbol{Z}^\top \boldsymbol{S}^{(g)} \boldsymbol{Z}) \tag{6.42}$$

$$\hat{w}_g = 0 \qquad \text{otherwise} \tag{6.43}$$

In fact this would be a too excessive sparse solution and too simple practically.

Other approaches on the same setting

The unique point of this problem setting is we have not only a set of features but also multiple sets of features. In machine learning, this setting is equivalent to so-called *multiview learning*, for which already numerous methods have been proposed under a variety of assumptions [14]. For example, a reasonable assumption is multiple views share a certain common set of features.

Another possible approach for the multiple sets of features is kernel learning. We can generate a kernel from each set of features, resulting in multiple kernels. *Multiple kernel learning* (MKL) is one matured approach in machine learning [36, 54].

Also in chapter 8 of this book, we consider a general situation in which we have multiple matrices sharing the same rows, starting with two matrices (in other words, each instance (row) is a set of multiple vectors), for which in Section 8.1.2, instances can be grouped (clustered). We show approaches for such general cases in Section 8.1.2.

6.2.2 Classification (Supervised Setting)

The problem of market segmentation is definitely clustering, because we do not have clear information on groups or classes of customers beforehand. However we may have some subtle knowledge on the assignment of customers to clusters, like that customers should be in the same cluster. This knowledge can be "constraints" rather than "prior knowledge". Of course if all instances are assigned to some classes or groups, the problem of assigning customers to groups is already solved, and so all instances cannot have clear information. Thus we can assume that only partial instances have some prior knowledge on cluster assignments. This problem setting is called *semisupervised learning* in machine learning. Again the problem would be clarified by defining the input and output of the problem:

Input:

1. (customers $\times$ features)-matrix $\boldsymbol{X}$. We have N customers, $\boldsymbol{x}_1, \ldots, \boldsymbol{x}_N$, i.e. $\boldsymbol{X} = (\boldsymbol{x}_1, \ldots, \boldsymbol{x}_N)^\top$. Let M be the size of features.

2. constraint. Below we consider two types of constraints, pairwise constraints $\boldsymbol{W}$ and matrix constraint $\boldsymbol{Y}$. See below for details.

Output:

1. $\boldsymbol{Z}$, segment assignment matrix, defined in Section 5.2.1, which has customers for rows and segments for columns.

Now let us consider this situation, which can be divided into the following two cases, i.e. pairwise and matrix constraints. We first briefly explain each of the two constraints, and then describe more detail on each further.

1. **Pairwise constraint:** We may already know that two customers should be in the same cluster. Or maybe a small number of people, like four people,

should be in the same cluster. We call this type of information *pairwise constraints*, which we incorporate into our problem setting. In fact a pairwise constraint is equivalent to an edge, connecting two nodes in a graph, in which each node corresponds to a unique customer. Thus this situation is equivalent to clustering vectors (customers) and at the same time clustering nodes in which edge connectivity must be followed by node clustering. That is, this problem setting is equivalent to the same situation as unsupervised learning "multiple data types" (particularly vectors and nodes in a graph) introduced in Section 5.2.3. This problem setting is called *semisupervised clustering* (with must-link constraints).

Let $\boldsymbol{W}$ be an adjacent matrix or given edge constraints, and $\boldsymbol{L}$ be graph Laplacian which can be derived from $\boldsymbol{W}$

2. **Matrix constraint:** The above situation is that we already know two or some small number of customers are in the same cluster. Now let us think that more complex knowledge are already obtained. It might be thought that it would not be so easy to define more complex knowledge than pairwise constraints. However the objective of disjoint clustering is to fill cluster assignment matrix $\boldsymbol{Z}$. Thus any complicated constraint cannot be more complicated than cluster assignment matrix $\boldsymbol{Z}$. Thus the assumption here is that we already know part of $\boldsymbol{Z}$ as $\boldsymbol{Y}$, which is a matrix with the same size as $\boldsymbol{Z}$ and in which true values of part of all elements are already assigned. Thus the problem setting is clustering, i.e. estimating $\boldsymbol{Z}$, while $\boldsymbol{Z}$ must be consistent with given $\boldsymbol{Y}$, which is again partial information on clustering customers.

We describe each of the above two cases of constraints in more detail below.

Pairwise Constraint: Semisupervised Clustering

Again this is the problem of semisupervised clustering with must-link constraints, which is described in Section 5.2.3. Thus we briefly review the procedure for solving this problem following Section 5.2.3.

First of all, this is clustering from given data $\boldsymbol{X}$, to estimate cluster assignment matrix $\boldsymbol{Z}$. If we use constrained K-means clustering and the squared distance for the difference between two instances (or an instance and the assigned cluster center), from given data $\boldsymbol{X}$, we first compute $\boldsymbol{E}$ with the following (i, j)-element $\boldsymbol{E}_{ij}$:

$$\boldsymbol{E}_{ij} = \frac{1}{2N} ||\boldsymbol{x}_i - \boldsymbol{x}_j||^2. \tag{6.44}$$

Then the problem can be formulated as the minimization problem as follows:

$$\min_{\boldsymbol{Z}} \operatorname{trace}(\boldsymbol{Z}^\top \boldsymbol{E} \boldsymbol{Z}). \tag{6.45}$$

Procedure 6.4: SEGMENTATION WITH PAIRWISE CONSTRAINTS.

Input: Matrix (customer × feature) $\boldsymbol{X}$; Pairwise constraints $\boldsymbol{W}$
Output: Segement assignment matrix $\boldsymbol{Z}$

Compute $\boldsymbol{E}$ and $\boldsymbol{L}$ from $\boldsymbol{X}$ and $\boldsymbol{W}$, respectively.;
Estimate $\hat{\boldsymbol{Z}}$ by solving the eigenvalue problem given by (6.50).;
Run `ClusteringPostProcessing(`$\hat{\boldsymbol{Z}}$`)`.;
Output finally estimated $\hat{\boldsymbol{Z}}$.;

On the other hand, given a graph of constraints, i.e. adjacency matrix $\boldsymbol{W}$, we can compute graph Laplacian $\boldsymbol{L}$, and the edge constraint can be the graph smoothness function as follows:

$$\mathrm{trace}(\boldsymbol{Z}^\top \boldsymbol{L}\boldsymbol{Z}) \leq \mathrm{Constant}. \tag{6.46}$$

Also we can consider the cluster size constraint, which has been used in many clustering approaches as follows:

$$\boldsymbol{Z}^\top \boldsymbol{Z} \;=\; \boldsymbol{D}, \tag{6.47}$$

where $\boldsymbol{D}$ is again a diagonal matrix, and diagonals specify cluster sizes.

Now we can formulate the semisupervised clustering problem as the following optimization problem:

$$\min_{\boldsymbol{Z}} \mathrm{trace}(\boldsymbol{Z}^\top \boldsymbol{E}\boldsymbol{Z}) \;\text{ subject to }\; \mathrm{trace}(\boldsymbol{Z}^\top \boldsymbol{L}\boldsymbol{Z}) \leq \mathrm{Constant} \;\text{ and }\; \boldsymbol{Z}^\top \boldsymbol{Z} = \boldsymbol{D}. \tag{6.48}$$

Then, by relaxing the binary value constraint of $\boldsymbol{Z}$ to allow real values, the Lagrangian can be given as follows:

$$L(\boldsymbol{Z}) = \mathrm{trace}(\boldsymbol{Z}^\top \boldsymbol{E}\boldsymbol{Z}) + \lambda_G \mathrm{trace}(\boldsymbol{Z}^\top \boldsymbol{L}\boldsymbol{Z}) + \lambda_C \mathrm{trace}(\boldsymbol{Z}^\top \boldsymbol{Z} - \boldsymbol{D}), \tag{6.49}$$

where λ_G and λ_C are hyperparameters.

Setting the partial derivative of the Lagrangian with respect to $\boldsymbol{Z}$ to zero, we can obtain the following eigenvalue problem:

$$(\boldsymbol{E} + \lambda_G \boldsymbol{L})\boldsymbol{Z} = \lambda_C \boldsymbol{Z} \tag{6.50}$$

Then solving this eigenvalue problem and running the postprocessing procedure (pseudocode) given by **Procedure 5.2** results in cluster assignment $\boldsymbol{Z}$. This procedure is summarized into the pseudocode in **Procedure 6.5**. The difference from the regular constrained K-means algorithm is using $(\boldsymbol{E} + \lambda_G \boldsymbol{L})$ instead of $\boldsymbol{E}$.

Procedure 6.5: SEGMENTATION WITH MATRIX CONSTRAINTS.

Input: (customer $\times$ feature)-matrix $\boldsymbol{X}$; Matrix constraints $\boldsymbol{Y}$
Output: Segment assignment matrix $\boldsymbol{Z}$

Compute $\boldsymbol{E}$ from $\boldsymbol{X}$, according to (6.44).;
Estimate $\hat{\boldsymbol{Z}}$ through (6.56) by using $\boldsymbol{E}$ and $\boldsymbol{Y}$.;
Output estimated $\hat{\boldsymbol{E}}$.;

Semisupervised Approach: Matrix Constraint

We assume that in given true cluster assignment $\boldsymbol{Y}$, only part of rows are true and others are unknown. That is, we assume that vectors $\boldsymbol{y}_i (i = 1, \ldots, Y)$ (corresponding to the first to the Y-th rows of $\boldsymbol{Y}$) are known, true cluster assignments. Then elements of the other rows are set at just zero.

We first consider the consistency between given true cluster labels $\boldsymbol{Y}$ and cluster assignment matrix $\boldsymbol{Z}$, as follows:

$$\min_{\boldsymbol{Z}} \text{trace}((\boldsymbol{Y} - \boldsymbol{Z})^\top (\boldsymbol{Y} - \boldsymbol{Z})) \tag{6.51}$$

Also we have data $\boldsymbol{X}$, and considering clustering over $\boldsymbol{X}$ with the squared distance between each instance and the corresponding cluster, we can compute $\boldsymbol{E}$, which is given by (6.44) and then the constraint from $\boldsymbol{X}$ can be given as follows:

$$\text{trace}(\boldsymbol{Z}^\top \boldsymbol{E} \boldsymbol{Z}) \leq \text{Constant}. \tag{6.52}$$

In addition, cluster assignment matrix $\boldsymbol{Z}$ itself has constraints on the cluster size:

$$\boldsymbol{Z}^\top \boldsymbol{Z} = \boldsymbol{D}. \tag{6.53}$$

Thus overall the problem can be formulated as the following optimization problem:

$$\begin{aligned} \min_{\boldsymbol{Z}} \quad & \text{trace}((\boldsymbol{Y} - \boldsymbol{Z})^\top (\boldsymbol{Y} - \boldsymbol{Z})) \\ \text{subject to} \quad & \text{trace}(\boldsymbol{Z}^\top \boldsymbol{E} \boldsymbol{Z}) \leq \text{Constant} \;\; \text{and} \;\; \boldsymbol{Z}^\top \boldsymbol{Z} = \boldsymbol{D}. \end{aligned} \tag{6.54}$$

Again by relaxing the binary constraint on $\boldsymbol{Z}$, we can set up the Lagrange multiplier:

$$L(\boldsymbol{Z}) = \text{trace}((\boldsymbol{Y} - \boldsymbol{Z})^\top (\boldsymbol{Y} - \boldsymbol{Z})) + \lambda_E \text{trace}(\boldsymbol{Z}^\top \boldsymbol{E} \boldsymbol{Z}) + \lambda_Z \text{trace}(\boldsymbol{Z}^\top \boldsymbol{Z} - \boldsymbol{D}), \tag{6.55}$$

where λ_E and λ_Z are hyperparameters.

From the Lagrangian, we can have the analytical solution of updating $\boldsymbol{Z}$ as follows:

$$\hat{\boldsymbol{Z}} = ((1 + \lambda_Z)\boldsymbol{I} + \lambda_E \boldsymbol{E})^{-1} \boldsymbol{Y}. \tag{6.56}$$

Practically, we estimate the $(Y+1)$-th to N-th rows of $\hat{\boldsymbol{Z}}$, by using the known part of $\boldsymbol{Z}$, i.e. $\boldsymbol{Y}$, following (6.56). This is equivalent to predicting cluster assignment of unknown instances (customers).

Note that this takes the same form as the analytical solution of label propagation in Section 5.2.2, where label propagation uses graph Laplacian $\boldsymbol{L}$ instead of $\boldsymbol{E}$, which can be computed from given data $\boldsymbol{X}$ here.

6.3 Market Targeting

6.3.1 Evaluating Segments

We have shown machine learning methods, which can automatically divide customers into segments by using customer features, as well as other information, particularly constraints on customer features and customers themselves. As explained in Section 3.2.1, the step next to market segmentation is market targeting, starting with segment evaluation. That is, each of the segments generated by market segmentation can be manually evaluated, and then one (or more) most appropriate segments can be selected. On the other hand, purely from a machine learning viewpoint, we attempt to evaluate the generated segments, by regarding them as clusters in clustering. We show methods to examine the quality of each cluster by using given data, and then we can compare the generated clusters (i.e. clusters) each other by using the results of the methods.

Examining Segment Quality: Internal Evaluation

We clarify the problem by showing the input and output of the setting:
Input:

1. (customers $\times$ features)-matrix $\boldsymbol{X}$. We have N customers, $\boldsymbol{x}_1, \ldots, \boldsymbol{x}_N$, i.e. $\boldsymbol{X} = (\boldsymbol{x}_1, \ldots, \boldsymbol{x}_N)^\mathsf{T}$. Let M be the size of features.

2. Segments $\boldsymbol{C} = \{C_1, \ldots, C_K\}$. We assume that these segments are obtained from $\boldsymbol{X}$.

Output:

1. Quality (C_k). function, which has segment C_k as input, outputs the quality score on cluster C_k.

Cluster quality can be evaluated by the distribution of instances in each cluster. That is, one idea is if feature values of instances in one cluster are concentrated more, this cluster would be better. Given cluster k, C_k, this can be simply checked by the mean distance between each instance and the representative vector of cluster k over all instances. In fact this idea is similar to the original clustering, by which we can consider the following metric, which is equivalent to the criterion of K-means clustering given in (5.30), as follows:

$$\text{Quality}(C_k) = \frac{1}{\sum_i \boldsymbol{Z}_{ik}} \sum_i \boldsymbol{Z}_{ik}\text{Dist}(\boldsymbol{x}_i, \boldsymbol{\mu}_k) \quad (6.57)$$

$$= \frac{1}{|C_k|} \sum_{i|\boldsymbol{x}_i \in C_k} \text{Dist}(\boldsymbol{x}_i, \boldsymbol{\mu}_k). \quad (6.58)$$

We can then use the squared distance for the distance, resulting in computing the variance of instances in each cluster:

$$\text{Quality}(C_k) = \frac{1}{|C_k|} \sum_{i|\boldsymbol{x}_i \in C_k} ||\boldsymbol{x}_i - \boldsymbol{\mu}_k||^2 \quad (6.59)$$

$$= \frac{1}{|C_k|} \sum_{i,j|\boldsymbol{x}_i \in C_k, \boldsymbol{x}_j \in C_k} ||\boldsymbol{x}_i - \boldsymbol{x}_j||^2. \quad (6.60)$$

Now let $\boldsymbol{W}^{(k)}$ be a binary matrix of (instances (customers) $\times$ instances), where the (i, j)-element is 1 if both the i-th instance and the j-th instance are in the same k-th cluster; otherwise zero.

In reality in $\boldsymbol{W}^{(k)}$, we can sort the instances according to clusters, for both the row and column sides, and the diagonal part corresponding to the k-th cluster have the elements filled by 1, and the other elements are zero. When we sort the instances as such, Fig. 6.1 (a) shows a schematic picture of $\boldsymbol{W}^{(k)}$, where the central gray part has the elements of one and in the other parts, all elements are zero. $\boldsymbol{W}^{(k)}$ shows the intra-cluster pairs of instances by 1 and the other pairs zero. Then by using $\boldsymbol{W}^{(k)}$, (6.60) can be further transformed into the following form:

$$\text{Quality}(C_k) = \frac{1}{|C_k|} \sum_{i,j|\boldsymbol{x}_i \in C_k, \boldsymbol{x}_j \in C_k} ||\boldsymbol{x}_i - \boldsymbol{x}_j||^2 \quad (6.61)$$

$$= \frac{1}{|C_k|} \sum_{i,j} \boldsymbol{W}^{(k)}_{ij} ||\boldsymbol{x}_i - \boldsymbol{x}_j||^2 \quad (6.62)$$

$$= \frac{1}{|C_k|} \sum_{i,j} \boldsymbol{W}^{(k)}_{ij} (2||\boldsymbol{x}_i||^2 - 2\boldsymbol{x}_i^\top \boldsymbol{x}_j) \quad (6.63)$$

$$= \frac{2}{|C_k|} \text{trace}(\boldsymbol{X}^\top (\boldsymbol{D}^{(k)} - \boldsymbol{W}^{(k)}) \boldsymbol{X}) \quad (6.64)$$

$$= \frac{2}{|C_k|} \text{trace}(\boldsymbol{X}^\top \boldsymbol{L}^{(k)} \boldsymbol{X}), \quad (6.65)$$

where $\boldsymbol{D}^{(k)}$ is a diagonal matrix and the (i, i)-th element is given by $\sum_j \boldsymbol{W}^{(k)}_{ij}$ and also $\boldsymbol{L}^{(k)}$ is given by $\boldsymbol{L}^{(k)} = \boldsymbol{D}^{(k)} - \boldsymbol{W}^{(k)}$.

In summary, this metric uses the clustering criterion given in (6.58) for each cluster, in terms of the sharpness/skewness of the instance distribution in the cluster. We can check this metric for each cluster, and finally we can pick up the most skewed cluster (with the minimum value of (6.58)). In particular, if we focus on the variance, we can use (6.65) for (6.58).

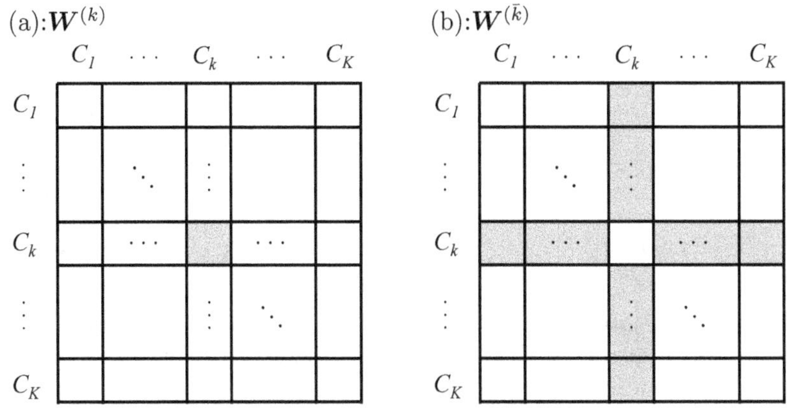

Figure 6.1: Schematic diagrams of two binary adjacency matrices: (a) In $\boldsymbol{W}^{(k)}$, the central gray part has elements between instances in cluster C_k and these elements take 1, while all other elements take zero, (b) In $\boldsymbol{W}^{(\bar{k})}$, the gray part has elements between instances in cluster C_k and instances in the other clusters. These elements take 1 and all other elements take zero.

Examining Segment Quality: Internal and External Evaluation

The above criterion to evaluate the generated cluster examines the distribution within each cluster. This metric for each cluster uses only instances in the corresponding cluster, which can be regarded as an *internal evaluation*. On the other hand, another idea would be, in order to compute some metric for one cluster, to use not only instances in the corresponding cluster but given other (or all) instances.

In this idea, under the same input and output as the internal evaluation, we can consider the distance between two different clusters. We can regard (6.60) as the *intra-cluster variance*, while we can consider the *inter-cluster variance*, which will be given as follows:

$$\text{Quality}(C_k) \quad = \quad \frac{1}{N - |C_k|} \sum_{i,j|\boldsymbol{x}_i \in C_k, \boldsymbol{x}_j \in C_{k'}, k \neq k'} ||\boldsymbol{x}_i - \boldsymbol{x}_j||^2. \qquad (6.66)$$

Now let $\boldsymbol{W}^{(\bar{k})}$ be a binary matrix of (instances $\times$ instances), where the (i, j)-element is 1 if the i-th instance is in the k-th cluster and the j-th instance is in another cluster, and vice versa, and the other (i, j)-element is zero. Fig. 6.1 (b) shows a schematic picture of $\boldsymbol{W}^{(\bar{k})}$, in which the grayed pair of instances are the elements with one, and the other pairs have zero. That is, $\boldsymbol{W}^{(\bar{k})}$ shows inter-cluster instance pairs (one instance in the k-th cluster and the other instance is in another cluster) by one, and the other pairs by zero. Then as we saw in the derivation of (6.65), by using $\boldsymbol{W}^{(\bar{k})}$, (6.66) can be equivalently written as follows:

$$\text{Quality}(C_k) = \frac{1}{N-|C_k|} \sum_{i,j|\boldsymbol{x}_i \in C_k, \boldsymbol{x}_j \in C_{k'}, k \neq k'} ||\boldsymbol{x}_i - \boldsymbol{x}_j||^2 \quad (6.67)$$

$$= \frac{2}{N-|C_k|} \text{trace}(\boldsymbol{X}^\top \boldsymbol{L}^{(\bar{k})} \boldsymbol{X}), \quad (6.68)$$

where $\boldsymbol{L}^{(\bar{k})} = \boldsymbol{D}^{(\bar{k})} - \boldsymbol{W}^{(\bar{k})}$ and $\boldsymbol{D}^{(\bar{k})}$ is a diagonal matrix and the (i,i)-th element is given by $\sum_j \boldsymbol{W}_{ij}^{(\bar{k})}$.

The idea of the internal evaluation was that the intra-cluster variance should be small, meaning that the distribution of instances in one cluster should be skewed. Thus in a reverse way, the inter-cluster variance should be large (comparing with intra-cluster variance), meaning that the distance between two arbitrary different clusters should be far.

Thus the criterion we can check can be given as follows:

$$\text{Quality}(C_k) = \frac{2}{|C_k|}\text{trace}(\boldsymbol{X}^\top \boldsymbol{L}^{(k)} \boldsymbol{X}) - \frac{2}{(N-|C_k|)}\text{trace}(\boldsymbol{X}^\top \boldsymbol{L}^{(\bar{k})} \boldsymbol{X}) \quad (6.69)$$

$$= \frac{2}{|C_k|}\text{trace}(\boldsymbol{X}^\top (\boldsymbol{L}^{(k)} - \lambda \boldsymbol{L}^{(\bar{k})}) \boldsymbol{X}), \quad (6.70)$$

where $\lambda = \frac{|C_k|}{(N-|C_k|)}$.

In particular, if we can assume the cluster size is equal over all clusters, we can ignore the normalization by the cluster size in (6.70) as follows:

$$\text{Quality}(C_k) = \text{trace}(\boldsymbol{X}^\top \boldsymbol{L}^{(k)} \boldsymbol{X}) - \text{trace}(\boldsymbol{X}^\top \boldsymbol{L}^{(\bar{k})} \boldsymbol{X}) \quad (6.71)$$

$$= \text{trace}(\boldsymbol{X}^\top (\boldsymbol{L}^{(k)} - \boldsymbol{L}^{(\bar{k})}) \boldsymbol{X}). \quad (6.72)$$

Thus the procedure to choose the most appropriate cluster is: we first compute (6.70) for each cluster and then choose the smallest one in terms of (6.70) out of all clusters as the best cluster (segment). This procedure is the same as the internal evaluation, while the above metric uses all instances to evaluate one cluster, instead of using only instances fallen into the corresponding cluster. **Procedure 6.6** shows a pseudocode of the above procedure for evaluating the segment (cluster) quality to choose one segment in terms of cluster skewness.

Evaluating Segment Attractivity from Customer Features

The last two metrics for evaluating the segments use only given $\boldsymbol{X}$ (or matrix of (customers $\times$ features)) except segments. Note here that each instance is a customer, by which each segment is a set of customers. We can consider another idea, in which we can evaluate each segment by using one particular feature of customers in the segment. For example, if the segment should be checked by its profitability, maybe we can focus on some related feature of customers, such as the *annual customer expense*. We can think of similar features, such as scale,

---

Procedure 6.6: SEGMENT QUALITY EVALUATION BY DATA SKEWNESS.

Input: (customers $\times$ features)-matrix $\boldsymbol{X}$; Segments (clusters) $C_1, \ldots, C_K$
Output: Ranked segments

1 **foreach** *segment* C_k **do**
2 　Compute the segment quality criterion by either (6.65) or (6.70).;
3 Rank $C_k (k = 1, \ldots, K)$ in the ascending order of the computed values.;

---

growth rate etc, for each segment. We can now here clarify the input and output of the current setting:

Input:

1. (customers $\times$ features)-matrix $\boldsymbol{X}$. We have N customers, $\boldsymbol{x}_1, \ldots, \boldsymbol{x}_N$, i.e. $\boldsymbol{X} = (\boldsymbol{x}_1, \ldots, \boldsymbol{x}_N)^\top$. Let M be the size of features.

2. Segments $\boldsymbol{C} = \{C_1, \ldots, C_K\}$. We assume that these segments are obtained from $\boldsymbol{X}$.

3. Feature value $\boldsymbol{v} = (v_1, \ldots, v_N)^\top$ for N customer. We assume that this feature is more important if v_i is larger.

Output:

1. Quality (C_k). function, which has segment C_k as input, outputs the quality score on cluster C_k.

Then the k-th segment can be evaluated by the sum of this feature value over all customers in this segment:

$$\text{Quality}(C_k) = \sum_{i|\boldsymbol{x}_i \in C_k} v_i \tag{6.73}$$

However instances are not necessarily equal in one segment, and the difference between instances should be considered: each instance can be weighed, for example, by the distance from the cluster center, which can be regarded as the similarity between an instance and the cluster center, $S(\boldsymbol{x}_i, \boldsymbol{\mu}_k)$. By using $S(\boldsymbol{x}_i, \boldsymbol{\mu}_k)$, we can then compute the quality of segment C_k as follows:

$$\text{Quality}(C_k) = \frac{\sum_{i|\boldsymbol{x}_i \in C_k} v_i S(\boldsymbol{x}_i, \boldsymbol{\mu}_k)}{\sum_{i|\boldsymbol{x}_i \in C_k} S(\boldsymbol{x}_i, \boldsymbol{\mu}_k)}, \tag{6.74}$$

which means that for example, for annual expense v_i of the i-th customer, Quality (C_k) is the weighted mean of the annual customer expense of segment k. Practically, we can consider (the simplest case, i.e.) cosine similarity: $S(\boldsymbol{x}_i, \boldsymbol{\mu}_k) = \boldsymbol{x}_i^\top \boldsymbol{\mu}_k$,

Procedure 6.7: SEGMENT QUALITY EVALUATION BY AN ARBITRARY FEATURE.

Input: (customer $\times$ feature)-matrix $\boldsymbol{X}$; Segments (clusters) $C_1, \ldots, C_K$; Feature $\boldsymbol{V}$

Output: Ranked segments

```
foreach segment C_k do
    Compute the segment quality criterion by (6.79).;
Rank all C_k (k = 1, ..., K) in the descending order of the computed values.;
```

and then the denominator of (6.74) can be written as follows:

$$\sum_{i|\boldsymbol{x}_i \in C_k} S(\boldsymbol{x}_i, \boldsymbol{\mu}_k) = \sum_{i|\boldsymbol{x}_i \in C_k} \boldsymbol{x}_i^\top \boldsymbol{\mu}_k \tag{6.75}$$

$$= \frac{1}{|C_k|} \sum_{i,j|\boldsymbol{x}_i \in C_k, \boldsymbol{x}_j \in C_k} \boldsymbol{x}_i^\top \boldsymbol{x}_j \quad (\Leftarrow \boldsymbol{\mu}_k = \frac{\sum_{j|\boldsymbol{x}_j \in C_k} \boldsymbol{x}_j}{|C_k|}) \tag{6.76}$$

The above term already appeared in the second term of (6.63), and we can write the similarity as follows:

$$\sum_{i|\boldsymbol{x}_i \in C_k} S(\boldsymbol{x}_i, \boldsymbol{\mu}_k) = \frac{1}{|C_k|} \mathrm{trace}(\boldsymbol{X}^\top \boldsymbol{W}^{(k)} \boldsymbol{X}). \tag{6.77}$$

Similarly let $\boldsymbol{V}$ be a diagonal matrix having the (i, i)-th element of v_i, the numerator of (6.74) can be written as follows:

$$\sum_{i|\boldsymbol{x}_i \in C_k} v_i S(\boldsymbol{x}_i, \boldsymbol{\mu}_k) = \frac{1}{|C_k|} \mathrm{trace}(\boldsymbol{X}^\top \boldsymbol{V} \boldsymbol{W}^{(k)} \boldsymbol{X}). \tag{6.78}$$

Thus overall the quality of the k-th segment can be examined as follows:

$$\mathrm{Quality}(C_k) = \frac{\mathrm{trace}(\boldsymbol{X}^\top \boldsymbol{V} \boldsymbol{W}^{(k)} \boldsymbol{X})}{\mathrm{trace}(\boldsymbol{X}^\top \boldsymbol{W}^{(k)} \boldsymbol{X})}. \tag{6.79}$$

Note that we can use any feature value for v_i, which is even not in the original matrix for generating segments. That is, $\boldsymbol{v}$, i.e. $\boldsymbol{V}$, can be given arbitrarily. **Procedure 6.7** shows a pseudocode of the above procedure for computing the quality of a segment, by using an arbitrary given feature.

6.3.2 Selecting Segments Using Customer Features

We have shown the metrics to examine the quality of each segment, which is given in (6.58), (6.70) and (6.79). Practically we can compute any of this metric for each of all segments, and sort segments by the resultant quality scores. This will provide helpful information for the company/firm/organization to choose the segment to go in.

6.4 Market Positioning

6.4.1 SWOT Analysis

As explained in Section 3.3.1, the SWOT analysis, an important step in market positioning, can help clearly understanding the environment under which the product is placed. Also the SWOT analysis has been used in a variety of other situations, such as considering the strategy of building a brand, etc. We here briefly review the SWOT analysis: under internal (controllable) and external (uncontrollable) environments, we think about good and bad effects, which can be summarized into strengths and weaknesses, respectively under the internal environment, and also into opportunities and threats, respectively, under the external environment. As shown in Table 3.2, the point of this analysis is to introduce practical aspects, such as *intellectual property (IP)* for strengths and *new technology* for opportunities, etc.

We might be able to raise such real, practical properties from data automatically. However, due to the nature that "opportunities" and "threats" are the aspects under external (uncontrollable) environments, it would be rather hard to detect them by only information we have for the segments. These aspects would need more complicated information. On the other hand, "strengths" and "weaknesses" might be still possible to be checked, because of their internal nature of environment. That is, we can compare the properties of the focused product with those of other products, since they are derived from internal and controllable environments. We then focus on the strengths and weaknesses of the SWOT analysis, and show that machine learning or some simpler statistical processing of data is useful to find the strengths and weaknesses of a product from given data. We first assume that a strength/weakness is one feature. In order to capture such a feature, we consider the following two steps:

1. **Data Collection:** We can first collect demographic data of a lot of products[2]. If we focus on some particular segment, we may just collect data of the products closely related with the focused segment. This step generates a data table of products for rows and features of those products for columns. Table 6.3 shows a schematic example of this data, and also substantially the same table was shown in Table 3.5 for the SWOT analysis.

2. **Feature Selection:** Again we assume that the strength or weakness of a product is one feature. Then if a feature is the strength (or weakness) of a product, the value of this product for the corresponding feature must be extremely high or low. We can then examine the distribution of values for each feature and select features, in which values of the focused product are extremely low or high (say top or bottom 1%), like *outliers*, in the distribution of the corresponding feature value. These features would become good

[2]When the SWOT analysis is applied to some product, we need collect product demographic data of a lot of competitors. This can be applied to the SWOT analysis on other cases, such as companies, services and brands. That is, we collect demographic data of a lot of companies, if the SWOT analysis is done for some company.

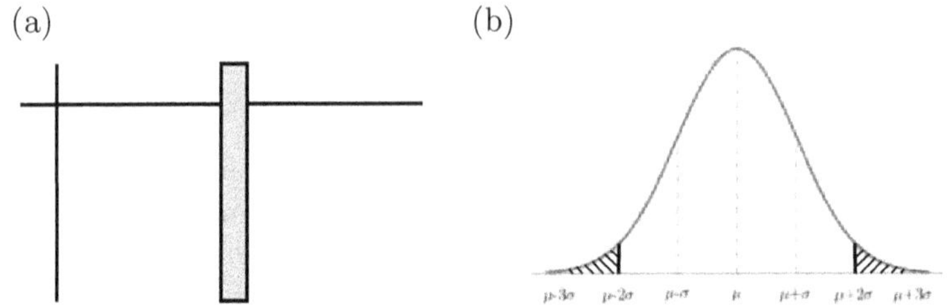

Figure 6.2: Possible two-step manner of automatics SWOT analysis: (a) We first collect data as a table of products (companies or services) with their features, where products are rows as instances and features are columns. Table 6.3 is an example of this table. (a) shows a schematic example of such data. Then in the second step, we examine each column (feature) of the matrix (table), as indicated by the dark-colored box in (a). Then the distribution of the values in each feature is examined, if the value of the corresponding product (company or service) is in the top or bottom (like the stripped parts in (b)), we keep this feature as a candidate of "strengths" or "weaknesses".

candidates for strengths or weaknesses. Fig. 6.2 shows a schematic picture of such selecting a possible feature, connected to "strength" and "weakness". Fig. 6.2 (a) shows a way of focusing one feature (column) of the product demographic table. Fig. 6.2 (b) shows a distribution of the focused feature, and check if the focused feature shows an extremely large or small value. For example, we can check if the feature value is in the stripped regions. If this value is so extreme, we keep this feature as a candidate for the "strength" or "weakness". However we note that we are unable to decide if this feature shows the strength, weakness, or having nothing to do with both. Thus this process just raises good candidates of strengths or weaknesses, and decision would be left.

Further Exploration

1. **Assigning strengths or weaknesses to the selected features** When we collect data, maybe features can be classified into positives or negatives as *labels*. For example, "annual sales" would be a positive feature, because it is better for the product if the value of this feature, i.e. the amount of annual sales, is larger. More generally for a positive feature, this feature can be a strength if the value of this feature is extremely large; otherwise a given feature is negative.

   We will then be able to assign the "strength" or "weakness" to each feature, if the value of this feature is extremely large or small. For example, if the value of a feature is extremely large (fallen into the right-hand side stripped region in Fig. 6.2 (b)) and this feature is "positive", this feature can be a "strength". Also if the value is extremely small and the feature

Procedure 6.8: DETECTING STRENGTHS/WEAKNESSES OF SWOT.

Input: (product × feature)-matrix $\boldsymbol{X}$; feature *labels* (positive/negative)
Output: strength set $\mathcal{S}$; weakness set $\mathcal{W}$

```
foreach feature k in X do
    if the value of feature k is extremely large/small then
        if the value is large and feature k is positive or small and negative
        then /* feature k is a ''strength''                    */
            Add feature k to S.;
        else /* feature k is ''weakness''                       */
            Add feature k to W.;
Output S and W.;
```

is negative, this feature is also "strength". On the other hand, the two other combinations: large and negative, and small and positive are both "weakness".

Thus, we can say that if we can first know if each feature is positive or negative, we can easily assign selected features (with extremely large or small values) to either of strength or weakness.

Assuming that each of all features is labeled as a positive or negative, **Procedure 6.8** shows a pseudocode of the procedure which we explained in the above for detecting strengths and weaknesses from given a matrix of products × features.

6.4.2 Generating Perceptual Map

As explained in Section 3.3.2, drawing a perceptual map is an important step to understand the market (or market segment) environment. Here we show several machine learning-based (or more simpler statistics-based) manners, which allow us to generate a perceptual map automatically, from the matrix of products (rows) with their features (columns). We note that this would be a general way, while recently a perceptual map (product positioning map) can be obtained by another way, e.g. using product reviews and text mining [75].

Notations

Let $\boldsymbol{X} = (\boldsymbol{x}_1, \ldots, \boldsymbol{x}_N)^\top = (\boldsymbol{f}_1, \ldots, \boldsymbol{f}_M)$ be a given matrix of products (rows) × features (columns), where $\boldsymbol{x}_i$ is the i-th instance, $\boldsymbol{f}_j$ is the j-th features, and $\boldsymbol{X}_{ij}$ be the entry corresponding the i-th product and j-th feature. Furthermore $x_i = \sum_j \boldsymbol{X}_{ij}$ and $f_j = \sum_i \boldsymbol{X}_{ij}$. Sample features were already shown in Table 3.4. Table 6.5 shows an example of data matrix $\boldsymbol{X}$ (on digital camera), using the features in Table 3.4.

Note that by general machine learning terms, $\boldsymbol{X}$ be a set of instances taking vectors, where each element is the value of the corresponding feature. Thus below we use the term "product" and "instance" interchangeably.

The feature values of $\boldsymbol{X}$ can be decided by experts, from demographic data (i.e. the specification of products) or from survey data (for example, each entry value can be the mean over the scores obtained by some survey).

Below we consider several problem settings for generating a perceptual map, which is an interesting and well-fitted application of machine learning. The generated map will be helpful for marketers to decide the strategy of positioning the product in market. If instances are other items, like companies, services or brands, visualizing such items would be also useful for marketers as well. Below we summarize the input and output of this section of generating perceptual map from given data

Input:

1. (products $\times$ features)-matrix $\boldsymbol{X}$. We have N products, $\boldsymbol{x}_1, \ldots, \boldsymbol{x}_N$, i.e. $\boldsymbol{X} = (\boldsymbol{x}_1, \ldots, \boldsymbol{x}_N)^\top$. Let M be the size of features.

2. (CSIP $\times$ feature)-matrix $\boldsymbol{\Phi}$. Note that $\boldsymbol{X}$ and $\boldsymbol{\Phi}$ share the same features. Also note that only $\boldsymbol{X}$ is used for learning a perceptual map and $\boldsymbol{\Phi}$ are just presented on the generated perceptual map, if necessary.

Output:

1. $\boldsymbol{D} = (\boldsymbol{d}_1, \ldots, \boldsymbol{d}_D)^\top$. The vector to be used to project given data $\boldsymbol{X}$ onto a perceptual map. A perceptual map is usually two-dimensional, and so $D = 2$ in general. Also in the generated perceptual map, the original input $\boldsymbol{X}$ are distributed over the 2D space, as $(\boldsymbol{X}\boldsymbol{d}_1, \boldsymbol{X}\boldsymbol{d}_2)$.

2D Space by the Selected Two Features

We assume that a perceptual map is two-dimensional (2D), and each dimension corresponds to one of given features. That is, the problem is to select two features out of given features. Then all instances (products) are distributed over the generated 2D space. The question here is the criterion to choose the two features. As briefly mentioned in Section 3.3.2, there exist two opposite ideas (directions) for this question:

Idea 1. Separate products as far as possible: Given products should be distributed as broadly as possible.

Idea 2. Position similar products as together as possible: Given products should be placed together if they are similar each other.

Idea 1 would be useful to show the difference of products, while **Idea 2** would be more useful for generating perceptual maps, since this clearly shows the similarity between products. Then these two ideas can be realized by the two following manners:

Procedure 6.9: PERCEPTUAL MAP BY TWO FEATURES WITH THE LARGEST VARIANCES.

Input: (product × feature)-matrix $\boldsymbol{X}$
Output: perceptual map

foreach *feature* $\boldsymbol{f}_k$ *in* $\boldsymbol{X}$ **do**
  Compute Var_k by using (6.80).;
Select $\boldsymbol{f}_k$ and $\boldsymbol{f}'_k$, which give the two largest p-values of (6.80) among all k.;
Generate the 2D space by letting $\boldsymbol{d}_1 = \boldsymbol{f}_k$ and $\boldsymbol{d}_2 = \boldsymbol{f}_{k'}$.;
Distribute $\boldsymbol{X}$ on the 2D space as $(\boldsymbol{X}\boldsymbol{d}_1, \boldsymbol{X}\boldsymbol{d}_2)$.;

Idea 1: → Select two features with the largest variances

In order to generate a 2D space, in which the given products are separated each other as far as possible, we can just focus on the feature which can distribute given instances as broadly as possible. This means that we can focus on the *variances* (more generally p-distances) of features, by which we can choose the two features which have the largest variances. Suppose that the range of values in each feature is already normalized, the p-distance of the k-th feature, Var_k can be written as follows:

$$\mathrm{Var}_k = \sum_i (x_{ik} - \mu_k)^p, \tag{6.80}$$

where $\mu_k = \frac{f_k}{N}$. In particular we can consider $p = 2$ for variances. Thus we can select the two features, each with the largest Var_k to generate the 2D space. **Procedure 6.9** shows a psuedocode of this procedure of selecting two features with the largest variances.

Idea 2: → Select two features which can keep the distance (by all features) between product pairs

The (p-)distance between two instances (products) i and i' by using all given features can be computed as follows:

$$\mathrm{Dist}(i, i') = \sum_j |x_{ij} - x_{i'j}|^p \tag{6.81}$$

Then the idea is that these distances over all product pairs should be kept, even after we pick up only two features for the perceptual map.

When we use only the k-th feature, the (p-)distance can be given as follows:

$$\mathrm{Dist}_k(i, i') = |x_{ik} - x_{i'k}|^p \tag{6.82}$$

Table 6.6: Notation and example for **Idea 2** of selecting two features.

q (product pairs)	$R(q)$	$R_k(q)$
q_1	1	3
q_2	2	1
q_3	3	5
q_4	4	2
...		

Thus the problem is to select a feature, so that the distance (6.82) by this feature, say the k-th feature, should be consistent with the distance obtained by (6.81) with all features. However these two distances have different scales, and the values from these two distances cannot be directly compared.

Then one possible, reasonable approach is that instead of actual distances, we can use ranking of product pairs, for example, sorted in the ascending order of the distances. That is, we first rank all product pairs by using the distances obtained by all features, i.e. (6.81), and we regard this ranking as the *gold standard ranking.* Similarly we can generate a ranking for the case of using only one feature. Then the problem becomes to detect the ranking (by only one feature), which is most consistent with the gold standard ranking. If we select the two most consistent features with the gold standard ranking, these features can be used to generate the 2D space.

Thus now the problem is how we can compute the similarity between an arbitrary given ranking and the gold standard ranking. For this problem, we can consider the following three measures to compute the similarity: 1) **Pointwise**, 2) **Pairwise** and 3) **Listwise** similarities, which will be explained below.

Notations: For simplicity we write q for the pair of product i and product i', i.e. $q = (i, i')$. Let $R(q)$ be the rank of product pair q in the gold standard ranking. For example, if $\text{Dist}(i, i')$ of products i and i' is smallest, $R(q)$ ($q = (i, i')$) is ranked top, i.e. $R(q) = 1$. Then similarly we can rank product pairs by using only the k-th feature, and let $R_k(q)$ be the rank of product pair q in this ranking (only the k-th feature is used). Table 6.6 shows an example of the notations.

Furthermore, for the pair of product pairs q and q', we define binary function $B_k(q, q')$, which shows the consistency of the ranking obtained by only the k-th feature with that by all features, in terms of the pair of two instances. That is, $B_k(q, q') = 1$ if $R(q) > R(q')$ and $R_k(q) > R_k(q')$, or $R(q) < R(q')$ and $R_k(q) < R_k(q')$; otherwise $B_k(q, q') = 0$. For example, if the ranking by only the k-th feature is totally the same as that by all features, $B_k(q, q') = 1$ for all pairs of q and q'.

Pointwise similarity

As shown in Table 6.6, each of the two columns, corresponding to $R(q)$ and $R_k(q)$, is a serious of integers, which can be regarded as a vector. We can then compare the two vectors, i.e. the gold standard ranking and one from the k-th feature, to compute the similarity. A simple approach is to compute the inner product or cosine similarity of two vectors. That is, we sum up product $R(q) \cdot R_k(q)$ over all q. We then normalize the inner product by the sizes of the two vectors, i.e. the size of $R(q)$ and that of $R_k(q)$, which is equivalent to (Pearson) correlation, given as follows:

$$\text{PointSim}_k = \frac{\sum_q R(q) \cdot R_k(q)}{\sqrt{\sum_q ||R(q)||^2} \cdot \sqrt{\sum_q ||R_k(q)||^2}} \tag{6.83}$$

We then check this similarity over all different k-th feature and pick up the two features with the largest similarities of (6.84). Note that if there are no tied instances in the two rankings, R and R_k, the sizes of the two rankings are the same, meaning that we do not have to think about normalization in (6.84):

$$\text{PointSim}_k = \sum_q R(q) \cdot R_k(q) \tag{6.84}$$

Pairwise similarity

The pairwise similarity between two rankings can be defined as the consistency of two pairs between the two rankings. For example, for pairs q and q', assuming that the rank of q is higher than that of q' in the gold standard ranking, if the rank of q is kept as higher than q in the ranking of using only the k-th feature, this pairwise consistency was kept; otherwise this became inconsistent.

That is, if $R(q) < R(q')$ and $R_k(q) < R_k(q')$ then the rank relationship between two pairs, (q, q'), was kept. Also if $R(q) > R(q')$ and $R_k(q) > R_k(q')$ then the rank relationship was kept Thus to compute the similarity, we can just count the case that the rank relationship is kept, which is equivalent to counting that $B_k(q, q') = 1$ for all possible pairs of q and q'. Thus the pairwise similarity is given as follows:

$$\text{PairSim}_k = \frac{\sum_{q,q'} B_k(q, q')}{\sum_{q,q'} 1} \tag{6.85}$$

Listwise similarity

In pointwise similarity (6.84), without changing the results, we can replace both $R(q)$ and $R_k(q)$ with $Q(q)$ $(= C - R(q))$ and $Q_k(q)$ $(= C - R_k(q))$, respectively, with some large constant C. Originally small $R(q)$ means a high rank, while $Q(q)$ should be large to be ranked high, meaning that $Q(q)$ is like *weights* (or *importance*), in the sense of larger $Q(q)$ for higher ranked q. This weight is indeed reasonable for higher ranked q, while for middle- or low-ranked q, this weight would not be so important. In other words, weight $Q(q)$ should be considered for only higher ranked q like top ten, and not for others. That is, weights should work for highly ranked q but the effect of weights should be more relaxed for other q.

Table 6.7: Q, discounted cumulative gain (DCG) and Q·DCG.

Rank	1	2	3	4	5	...
$Q(q)$ ($R(q)$ = Rank)	5	4	3	2	1	...
$Q_k(q)$ ($R(q)$ = Rank)	3	1	5	2	4	...
$\mathrm{DCG}_{\mathrm{Rank}} = 1/(\log_2(\mathrm{Rank}+1))$	1	0.631	0.5	0.431	0.387	...
$Q(q)\cdot \mathrm{DCG}_{\mathrm{Rank}}$	5	2.52	1.5	0.862	0.387	...
$Q_k(q)\cdot \mathrm{DCG}_{\mathrm{Rank}}$	3	0.631	2.5	0.862	1.548	...

One possible approach to implement this idea is *discounted cumulative gain (DCG)*, which is defined for rank r, as follows:

$$\mathrm{DCG}_r = \frac{1}{\log_2(r+1)}. \tag{6.86}$$

The behavior of DCG would be able to be understood by examples more. Table 6.7 shows actual values of DCG (fourth row) for rank r ($r = 1, 2, 3, ...$) (from left to right). As shown in this table, different from the regular rank (first row), which has always an equal interval, DCG (fourth row) has no more big difference for lower ranked instances. Also this table shows $Q(q)$ (second row) for the gold standard ranking which just shows the reverse order of Rank, and example values of $Q_k(q)$ (third row). Finally we can compute the similarity score for the case of using only the k-th feature, as follows:

$$\mathrm{ListSim}_k = \frac{\sum_q Q_k(q)\cdot \mathrm{DCG}_{R(q)}}{\sum_q Q(q)\cdot \mathrm{DCG}_{R(q)}}. \tag{6.87}$$

This similarity is equivalent to *normalized discounted cumulative gain (nDCG)*.

Procedure 6.10 shows an example procedure of the selection of two features (out of all given features) to generate a perceptual map. In this procedure we note that we can use any of the three similarities we raised in the above, i.e. the pointwise, pairwise and listwise similarities.

Notes on the similarities raised for Idea 2

Connection to "Learning to Rank" We have introduced three similarities: pointwise, pairwise and listwise similarities. These three similarities correspond to the ideas of three representative metrics, which have been developed consecutively in the history of a relatively new machine learning paradigm, *learning to rank* [63, 65].

Complexity of pairwise similarity Among the three similarities, for each feature (say the k-th feature), pairwise similarities have to consider the pair of products, while the other two similarities examine each product independently. Thus computing the pairwise similarity has a larger time complexity (the square time complexity, while the the other two have just the linear

Procedure 6.10: PERCEPTUAL MAP BY TWO FEATURES CONSISTENT WITH USING ALL FEATURES.

Input: (products $\times$ features)-matrix $\boldsymbol{X}$
Output: perceptual map

Compute distances over all product pairs from $\boldsymbol{X}$ by using (6.81).;
foreach *feature* $\boldsymbol{f}_k$ *in* $\boldsymbol{X}$ **do**
    Compute distances over all product pairs from $\boldsymbol{X}$ by using (6.82).;
    Compute similarity Sim_k by using either of (6.84), (6.85) or (6.87).;
Select $\boldsymbol{f}_k$ and $\boldsymbol{f}_{k'}$, which provide the two largest Sim_k among all $\boldsymbol{f}_i (i = 1, \dots, M)$, as $\boldsymbol{d}_1$ and $\boldsymbol{d}_2$, respectively.;
Generate the 2D space by using $\boldsymbol{d}_1$ and $\boldsymbol{d}_2$.;
Distribute $\boldsymbol{X}$ on the 2D space as $(\boldsymbol{X}\boldsymbol{d}_1, \boldsymbol{X}\boldsymbol{d}_2)$.;

time complexity), which makes applying the pairwise similarity to the large-scale data hard. However, instances of given $\boldsymbol{X}$ are products, and the size of products would not be so large (if that is too large, all products cannot be visualized over the 2D space). Thus in general this square complexity would not become a big issue.

Generating Customer Segment Ideal Points (CSIPs)

In Section 3.3.2, we show one perceptual map with not only products (written as competitors in Section 3.3.2) but also customer segment ideal points (CSIPs), where each CSIP represents the corresponding segment. For example, Fig. 3.4 shows one schematic example of showing competitors as well as CSIPs on a 2D perceptual map. If we regard each CSIP as the representative point of the corresponding segment, i.e. cluster in machine learning, a simple idea is to assign the cluster (segment) center to the corresponding CSIP $\boldsymbol{\Phi}_k$, as follows:

$$\phi_k = \frac{1}{|C_k|} \sum_{i|\boldsymbol{x}_i \in C_k} \boldsymbol{x}_i, \qquad (6.88)$$

where $\boldsymbol{x}_i$ is a vector corresponding to the i-th customer.

In marketing, however, it is sometimes thought that there must exist multiple CSIPs for each segment [25]. In other words, only one CSIP would not be good enough to represent one segment. In fact, this would be reasonable, because we can use different features between generating segments and also generating a perceptual map, which might generate more than one CSIPs even for segment.

Table 6.8 shows the way to generate multiple CSIPs for one segment schematically. Starting with the original survey data (top of Table 6.8), we can generate the matrix of customers $\times$ features (middle of Table 6.8). Also these customers can be grouped according to segments generated in market segmentation (bottom of Table 6.8). Then in fact features in the survey data are not necessarily the same as the features used for market segmentation, and so within each segment,

Table 6.8: (top) Original survey data, which is the same as the top of Tables 6.2 and 6.4. (middle) (Customers × Features)-table obtained by, for each customer, averaging feature values over products. (bottom) Customers can be grouped and reordered by segments obtained in market segmentation. Furthermore, since features which are used for generating segments are different from those in these tables, we can do clustering customers within each segment by using these features. This clustering generates cluster centers, which correspond to customer segment ideal points (CSIPs), in each segment. That is, we can have multiple CSIPs for each segment.

		Feature 1	Feature 2	Feature 3	...
Customer1	Product A				
	Product B				
	...				
Customer2	Product A				
	Product B				
	...				
Customer3	Product A				
	Product B				
	...				

⇓

	Feature 1	Feature 2	Feature 3	...
Customer1				
Customer2				
...				

⇓

		Feature 1	Feature 2	Feature 3	...
Segment i	Customer3				
	Customer5				
	...				
Segment ii	Customer1				
	Customer7				
	...				
Segment iii	Customer2				
	Customer6				
	...				

Procedure 6.11: GENERATING MULTIPLE CONSUMER SEGMENT IDEAL POINTS FOR EACH SEGMENT.

Input: (product $\times$ feature)-matrix $\boldsymbol{X}$; Segment assignment matrix $\boldsymbol{Z}$
Output: multiple CSIPs

```
Generate a matrix of customers separated by segments, like the bottom of
  Table 6.8, from both X and Z.;
foreach segment of Z do
    Run a clustering algorithm, say constrained K-means clustering, over
      customers in the segment by using features of X.;
    Provide CSIPs with K cluster centers.;
Output all CSIPs.;
```

---

we can run a clustering algorithm over customers again by using features derived from the survey data. Note that we can run any type of clustering algorithm, such as constrained k-means clustering. Then by using the resultant clusters of each segment, we can provide each CSIP in a segment with one cluster center of the corresponding segment, by which we can have multiple CSIPs within each segment. **Procedure 6.11** is a pseudocode of this procedure of generating multiple CSIPs for each segment, by using the features of survey data which are different from those for market segmentation.

2D Space by Multiple (All) Features

We can generate each dimension of a 2D perceptual map by using multiple (all) features and then distribute products over the 2D space. In other words, each dimension is not one single feature but some combination of all (or part) of features. Then the question is the criterion to generate the combination of features for generating the two dimensions of a perceptual map. Again we can have the following two ideas, which are already shown in the case that the dimensions are generated by using only two features:

Idea 1. Separate products as far as possible: Given products should be distributed as broadly as possible.

Idea 2. Position similar products as together as possible: Given products should be placed together if they are similar each other.

Also again about these two ideas, **Idea 1** would be useful to show the difference of products, while **Idea 2** would be more useful for generating perceptual maps, since this clearly shows the similarity between products.

We use the same notation, where input data is $\boldsymbol{X}$ in which rows are products (instances) and columns are their features.

Idea 1: $\rightarrow$ Principal Component Analysis (PCA)

If we focus on **Idea 1**, one possible approach would be principal component analysis (PCA), which we already described in feature learning in Section 5.3.2. The idea of PCA is to project the data onto an one-dimensional (1D) space so that the variance on the 1D space should be maximized (Fig. 5.5). PCA maximizes the broadness (variance) of the distribution of the plots on the 1D space, which is obtained by the projection. This idea of PCA is exactly consistent with the above **Idea 1**. Below we briefly review the procedure of PCA in Section 5.3.2 once again.

We can first write the projection on the 1D space as $\boldsymbol{X}\boldsymbol{w}$, where $\boldsymbol{w}$ is the parameter (linear coefficient) to be estimated from data. Suppose that $\boldsymbol{X}\boldsymbol{w}$ is already normalized so that the mean is zero, we can compute the variance as follows:

$$(\boldsymbol{X}\boldsymbol{w})^\top \boldsymbol{X}\boldsymbol{w} = \boldsymbol{w}^\top \boldsymbol{X}^\top \boldsymbol{X}\boldsymbol{w}. \tag{6.89}$$

The objective function is to maximize this $\boldsymbol{w}^\top \boldsymbol{X}^\top \boldsymbol{X}\boldsymbol{w}$, while $\boldsymbol{w}$ should be regularized, for example, as follows:

$$||\boldsymbol{w}||^2 = \boldsymbol{w}^\top \boldsymbol{w} = \text{Constant}. \tag{6.90}$$

From these two, we can formulate an optimization problem, for which Lagrangian $L(\boldsymbol{w})$ of this problem can be written as follows:

$$L(\boldsymbol{w}) = \boldsymbol{w}^\top \boldsymbol{X}^\top \boldsymbol{X}\boldsymbol{w} - \lambda(\boldsymbol{w}^\top \boldsymbol{w} - \text{Constant}), \tag{6.91}$$

where λ is a Lagrange multiplier.

We set the gradient of Lagrangian $L(\boldsymbol{w})$ with respect to $\boldsymbol{w}$ equal to zero, and this results in an eigenvalue problem, as follows:

$$\frac{dL(\boldsymbol{w})}{d\boldsymbol{w}} = 0 \qquad \Rightarrow \qquad \boldsymbol{X}^\top \boldsymbol{X}\boldsymbol{w} = \lambda \boldsymbol{w}. \tag{6.92}$$

After solving this eigenvalue problem, the first and second eigenvectors, $\boldsymbol{w}_1$ and $\boldsymbol{w}_2$, respectively, are $\boldsymbol{d}_1$ and $\boldsymbol{d}_2$, respectively, which generate the 2D space and perceptual map over which the products are plotted, and this can be obtained by the following:

$$\boldsymbol{X}\boldsymbol{w}_1 \ (= \boldsymbol{X}\boldsymbol{d}_1), \qquad \text{and} \qquad \boldsymbol{X}\boldsymbol{w}_2 \ (= \boldsymbol{X}\boldsymbol{d}_2). \tag{6.93}$$

Also if we have a newly given instance $\boldsymbol{x}_{\text{new}}$, we can compute the following values to put this instance on the 2D space:

$$\boldsymbol{x}_{\text{new}}^\top \boldsymbol{d}_1, \qquad \text{and} \qquad \boldsymbol{x}_{\text{new}}^\top \boldsymbol{d}_2. \tag{6.94}$$

That is, we can distribute given instances (products) over the 2D space, generated by the above two axes.

Modification of PCA

Procedure 6.12: PRINCIPAL COMPONENT ANALYSIS FOR GENERATING A PERCEPTUAL MAP.

Input: (product × feature)-matrix $\boldsymbol{X}$
Output: perceptual map

1 Run eigenvalue decomposition over $\boldsymbol{X}$ (6.92) to have two top eigenvectors: $\boldsymbol{w}_1$ and $\boldsymbol{w}_2$, which are used as $\boldsymbol{d}_1$ and $\boldsymbol{d}_2$, respectively.;
2 Products are projected onto 2D space by $\boldsymbol{X}\boldsymbol{d}_1$ and $\boldsymbol{X}\boldsymbol{d}_2$.;

Modification 1. Using kernel function We may want to incorporate some prior (background) knowledge on the products, for example, similarity between products. One possible approach would be an extension to PCA under kernel learning, i.e. kernel principal component analysis (KPCA), by using kernels for the product similarities. We now describe this extension (see more detail in [70]).

First we again review PCA. Given $\boldsymbol{X}$, the optimization function by PCA was given as follows:

$$\min_{\boldsymbol{w}} \boldsymbol{w}^\top \boldsymbol{X}^\top \boldsymbol{X} \boldsymbol{w} \quad \text{subject to} \quad \boldsymbol{w}^\top \boldsymbol{w} = \text{Constant}, \tag{6.95}$$

which was given in (5.85) already.

We assume that parameter $\boldsymbol{w}$ can be a linear combination of input $\boldsymbol{X}$:

$$\boldsymbol{w} = \boldsymbol{X}^\top \boldsymbol{\beta}. \tag{6.96}$$

and substitute this for (6.95). Then (6.95) can be reformulated as the following optimization problem:

$$\max_{\boldsymbol{\beta}} \boldsymbol{\beta}^\top \boldsymbol{X}\boldsymbol{X}^\top \boldsymbol{X}\boldsymbol{X}^\top \boldsymbol{\beta} \quad \text{subject to} \quad \boldsymbol{\beta}^\top \boldsymbol{X}\boldsymbol{X}^\top \boldsymbol{\beta} = \text{Const}. \tag{6.97}$$

We can then define the Lagrange function for this optimization problem as follows:

$$L(\boldsymbol{\beta}) = \boldsymbol{\beta}^\top \boldsymbol{X}\boldsymbol{X}^\top \boldsymbol{X}\boldsymbol{X}^\top \boldsymbol{\beta} - \lambda \boldsymbol{\beta}^\top \boldsymbol{X}\boldsymbol{X}^\top \boldsymbol{\beta}. \tag{6.98}$$

By setting the derivative of the Lagrange function with respect to $\boldsymbol{\beta}$ equal to zero, we obtain the following eigenvalue problem:

$$\frac{dL(\boldsymbol{\beta})}{d\boldsymbol{\beta}} = 0 \qquad \Rightarrow \qquad \boldsymbol{X}\boldsymbol{X}^\top \boldsymbol{\beta} = \lambda \boldsymbol{\beta}. \tag{6.99}$$

We replace $\boldsymbol{X}\boldsymbol{X}^\top$ in the right-hand size of (6.99) with kernel function $\boldsymbol{K}$:

$$\boldsymbol{X}\boldsymbol{X}^\top \qquad \Rightarrow \qquad \boldsymbol{K}. \tag{6.100}$$

Finally we can see that simply $\boldsymbol{\beta}$ can be estimated by solving the eigenvalue decomposition of the kernel function:

$$\boldsymbol{K}\boldsymbol{\beta} = \lambda \boldsymbol{\beta}. \tag{6.101}$$

Procedure 6.13: KERNEL PRINCIPAL COMPONENT ANALYSIS FOR GENERATING A PERCEPTUAL MAP.

Input: (product $\times$ product)-kernel: $\boldsymbol{K}$
Output: Perceptual map

1 Run eigenvalue decomposition over $\boldsymbol{K}$ (6.101) to have two top eigenvectors: $\boldsymbol{\beta}_1$ and $\boldsymbol{\beta}_2$, as $\boldsymbol{d}_1$ and $\boldsymbol{d}_2$, respectively.;
2 Products are projected onto 2D space by $\boldsymbol{K}\boldsymbol{d}_1$ and $\boldsymbol{K}\boldsymbol{d}_2$.;

We then consider the projection of $\boldsymbol{X}$ into the obtained one-dimensional space: originally in PCA, instances $\boldsymbol{X}$ are projected into one dimensional vector by using $\boldsymbol{w}$:

$$\boldsymbol{X}\boldsymbol{w} \tag{6.102}$$

We then use both (6.96) and (6.102), to project into one dimensional vector using only $\boldsymbol{\beta}$, instead of $\boldsymbol{w}$:

$$\boldsymbol{X}\boldsymbol{w} = \boldsymbol{X}\boldsymbol{X}^\top\boldsymbol{\beta} \tag{6.103}$$
$$= \boldsymbol{K}\boldsymbol{\beta}. \tag{6.104}$$

Thus after solving the eigenvalue problem of (6.101), the first and second eigenvectors, $\boldsymbol{\beta}_1$ and $\boldsymbol{\beta}_2$, respectively, are used as $\boldsymbol{d}_1$ and $\boldsymbol{d}_2$, respectively, to generate the 2D space and perceptual map over which the products are plotted, as follows:

$$\boldsymbol{K}\boldsymbol{\beta}_1 \ (= \boldsymbol{K}\boldsymbol{d}_1), \quad \text{and} \quad \boldsymbol{K}\boldsymbol{\beta}_2 \ (= \boldsymbol{K}\boldsymbol{d}_2). \tag{6.105}$$

Also if we have a newly given instance $\boldsymbol{x}_{\text{new}}$, we first compute the similarity (element of the kernel matrix) against each of other instances in kernel function $\boldsymbol{K}$, resulting in generating vector $\boldsymbol{k}_{\text{new}}$. Then we compute the following values to put this instance on the 2D space:

$$\boldsymbol{k}_{\text{new}}^\top\boldsymbol{d}_1, \quad \text{and} \quad \boldsymbol{k}_{\text{new}}^\top\boldsymbol{d}_2. \tag{6.106}$$

Procedure 6.13 shows a pseudocode of the procedure to have a perceptual map generated by kernel PCA. In this algorithm we just focus on only two top eigenvectors, while general kernel PCA can provide all eigenvectors, which might be useful to project given data onto a smaller dimensional space.

Idea 2: $\rightarrow$ Correspondence Analysis (CA)

In order to implement **Idea 2**, one possible approach would be to use *correspondence analysis* (CA), which has been traditionally considered as a standard

Table 6.9: Residual r_{ij} shows the correlation between product i and feature j.

Residual	Correlations
$r_{ij} > 0$	Positive correlation
$r_{ij} = 0$	Independence
$r_{ij} < 0$	Negative correlation

method for generating perceptual maps in marketing [19, 43, 10]. CA has a step of preprocessing data (this preprocessing is somewhat too emphasized when CA has been explained in general), while the key point of CA would be the difference from PCA. Below we will explain the difference between PCA and CA. First PCA makes the distribution of given instances (or products) as diversely as possible, by focusing on the variance matrix of instances (which is the inner–product (similarity) among features), i.e. $\boldsymbol{X}^\top \boldsymbol{X}$, which is then run by eigenvalue decomposition to have eigenvectors, resulting in the two axes of the 2D space. On the other hand, CA focuses on the inner–product (or similarity) among products, i.e. $\boldsymbol{X}\boldsymbol{X}^\top$, which is then used to have the two axes of the 2D space by eigenvalue decomposition. Thus simply speaking, both PCA and CA are eigenvalue decomposition, while this decomposition is over $\boldsymbol{X}^\top \boldsymbol{X}$ in PCA and $\boldsymbol{X}\boldsymbol{X}^\top$ in CA. Below we will explain CA, which has the following two steps:

1. **Preprocessing data** In this step, we compute the following matrix $\boldsymbol{R}$ (in which (i, j)-element is $\boldsymbol{R}_{ij}$) from the input matrix $\boldsymbol{X}$, as follows:

$$\boldsymbol{R}_{ij} = \frac{\boldsymbol{X}_{ij} - e_{ij}}{\sqrt{e_{ij}}}, \tag{6.107}$$

where $e_{ij} = x_i \cdot f_j$. That is, e_{ij} is the *expectation value* of the i-th product and j-th feature, in terms that the i-th product and j-th feature happen at the same time, assuming that they are independent. Here $\boldsymbol{R}_{ij}$ is called the *(standardized) residual* of $\boldsymbol{X}_{ij}$. In fact $\boldsymbol{R}_{ij}$ shows the correlation between the i-th product and j-th feature. Table 6.9 summarizes the relationships between the value of $\boldsymbol{R}_{ij}$ and the correlation between the i-th product and j-th feature:

We use r_{ij} as preprocessed data for further computation[3]. We note that this preprocessing follows the derivation of χ^2 statistic, which can be computed as follows:

$$\chi^2\text{statistic} = \sum_i^N \sum_j^M \frac{(\boldsymbol{X}_{ij} - e_{ij})^2}{e_{ij}} \tag{6.109}$$

[3]Some variations exist for the preprocessing step. For example, instead of the residual, the following simpler normalization is also used:

$$\boldsymbol{R}_{ij} = \frac{\boldsymbol{x}_{ij}}{e_{ij}}. \tag{6.108}$$

Procedure 6.14: PERCEPTUAL MAP BY CORRESPONDENCE ANALYSIS.

Input: (products × features)-matrix $\boldsymbol{X}$
Output: perceptual map

1 Preprocess data by transforming $\boldsymbol{X}$ into $\boldsymbol{R}$ by using (6.107).;
2 Run singular value decomposition over $\boldsymbol{R}$ as shown in (6.110).;
3 Generate the 2D space by using the first two columns of $\boldsymbol{U}$, i.e. eigenvectors $\boldsymbol{u}_1$ and $\boldsymbol{u}_2$, which are $\boldsymbol{d}_1$ and $\boldsymbol{d}_2$, respectively.;
4 Project products onto 2D space by $\boldsymbol{R}\boldsymbol{R}^\top\boldsymbol{u}_1$ and $\boldsymbol{R}\boldsymbol{R}^\top\boldsymbol{u}_2$.;

Thus we can say that this preprocessing step transforms the original input data into the data which follows χ^2 distribution. Also note that χ^2 test is known as an approximation of Fisher's exact test, which assumes the hypergeometric distribution of given instances.

2. Similarity data and singular value decomposition We then compute the *singular value decomposition* (SVD) of $\boldsymbol{R}$ as follows:

$$\boldsymbol{R} = \boldsymbol{U}\boldsymbol{\Lambda}\boldsymbol{V}^\top, \tag{6.110}$$

where $\boldsymbol{\Lambda}$ is a diagonal matrix and $\boldsymbol{U}$ and $\boldsymbol{V}$ are orthogonal matrices, satisfying:

$$\boldsymbol{U}\boldsymbol{U}^\top = \boldsymbol{I}, \quad \boldsymbol{V}\boldsymbol{V}^\top = \boldsymbol{I}, \tag{6.111}$$

where $\boldsymbol{I}$ is an identity matrix.

Readers can see Appendix (Page 344) of [70] for further details of SVD. We then focus on $\boldsymbol{U}$, and use the first two columns (vectors) of $\boldsymbol{U}$ to generate the 2D space of the perceptual map, in which products are plotted.

There exist some modifications (normalization) against just using $\boldsymbol{U}$. For example, instead of $\boldsymbol{U}$, $\boldsymbol{U}\boldsymbol{\Lambda}$ can be used. Furthermore $\boldsymbol{U}\boldsymbol{\Lambda}$ can be further normalized to $\boldsymbol{D}\boldsymbol{U}\boldsymbol{\Lambda}$, where $\boldsymbol{D}$ is a diagonal matrix with d_{ii} for the i-th row and the i-th column and d_{ii} is $\frac{1}{\sqrt{x_i}}$.

Entirely **Procedure 6.14** summarizes the above two step procedure of CA into a pseudocode.

Interpretation

Interpretation 1. Equivalence to the eigenvalue problem of $\boldsymbol{R}\boldsymbol{R}^\top$ This is rather obvious due to the nature of SVD. We can compute the inner product matrix of $\boldsymbol{R}$, i.e. $\boldsymbol{R}\boldsymbol{R}^\top$, as follows:

$$\begin{aligned} \boldsymbol{R}\boldsymbol{R}^\top &= \boldsymbol{U}\boldsymbol{\Lambda}\boldsymbol{V}^\top(\boldsymbol{U}\boldsymbol{\Lambda}\boldsymbol{V}^\top)^\top = \boldsymbol{U}\boldsymbol{\Lambda}\boldsymbol{V}^\top(\boldsymbol{V}\boldsymbol{\Lambda}^\top\boldsymbol{U}^\top) && (6.112) \\ &= \boldsymbol{U}\boldsymbol{\Lambda}\boldsymbol{\Lambda}^\top\boldsymbol{U}^\top = \boldsymbol{U}\boldsymbol{\Lambda}^2\boldsymbol{U}^\top. && (6.113) \end{aligned}$$

Since $\boldsymbol{U}$ is orthogonal, we can have the following:

$$\boldsymbol{R}\boldsymbol{R}^{\mathsf{T}}\boldsymbol{U} = \boldsymbol{\Lambda}^2\boldsymbol{U} \tag{6.114}$$

This is exactly the formulation of eigenvalue decomposition of $\boldsymbol{R}\boldsymbol{R}^{\mathsf{T}}$, where $\boldsymbol{U}$ is eigenvectors and $\boldsymbol{\Lambda}^2$ has eigenvalues for diagonals.

Thus this means that **Procedure 6.14** can be written in the following five steps, by using eigenvalue decomposition instead of singular value decomposition:

Step 1. Compute residuals $\boldsymbol{R}$ from input $\boldsymbol{X}$.

Step 2. Compute inner-product matrix $\boldsymbol{R}\boldsymbol{R}^{\mathsf{T}}$.

Step 3. Run eigenvalue decomposition over $\boldsymbol{R}\boldsymbol{R}^{\mathsf{T}}$.

Step 4. Use the first two columns of $\boldsymbol{U}$: $\boldsymbol{u}_1$ for $\boldsymbol{d}_1$ and $\boldsymbol{u}_2$ for $\boldsymbol{d}_2$.

Step 5. Project data by using $\boldsymbol{R}$, $\boldsymbol{d}_1$ and $\boldsymbol{d}_2$ into 2D space $(\boldsymbol{R}\boldsymbol{R}^{\mathsf{T}}\boldsymbol{d}_1, \boldsymbol{R}\boldsymbol{R}^{\mathsf{T}}\boldsymbol{d}_2)$.

In fact the difference is, instead of $\boldsymbol{R}$, we compute $\boldsymbol{R}\boldsymbol{R}^{\mathsf{T}}$ to run eigenvalue decomposition over that. Except that, the procedure is totally kept as the same.

Interpretation 2. Equivalence to clustering

Again $\boldsymbol{R}\boldsymbol{R}^{\mathsf{T}}$ is a $(\boldsymbol{N} \times \boldsymbol{N})$ similarity matrix of instances (products) in $\boldsymbol{X}$. This matrix can be regarded as an adjacency matrix of the instances in $\boldsymbol{X}$. That is, $\boldsymbol{R}\boldsymbol{R}^{\mathsf{T}}$ can be regarded as a graph with N nodes, where edges show similarity between corresponding nodes.

Then we can consider binary *cluster assignment matrix* $\boldsymbol{Z}$ with N rows for products and K columns for clusters. Then if $\boldsymbol{Z}_{ik}$ is 1, the i-th product is in the k-th cluster; otherwise the i-th product is not in the k-th cluster.

We consider the problem to estimate $\boldsymbol{Z}$ from the graph, i.e. $\boldsymbol{R}\boldsymbol{R}^{\mathsf{T}}$. We can compute the consistency/smoothness of $\boldsymbol{Z}$ over the graph as follows[4]:

$$\boldsymbol{Z}^{\mathsf{T}}\boldsymbol{L}\boldsymbol{Z}, \tag{6.115}$$

where $\boldsymbol{L}$ is the graph Laplacian of the adjacency matrix and given as follows:

$$\boldsymbol{L} \quad = \quad \boldsymbol{D} - \boldsymbol{R}\boldsymbol{R}^{\mathsf{T}}, \tag{6.116}$$

where $\boldsymbol{D}$ is a diagonal matrix and the i-th diagonal $\boldsymbol{D}_{ii}$ is the sum over all column values of the i-th row of $\boldsymbol{R}\boldsymbol{R}^{\mathsf{T}}$.

In order to solve the problem we addressed, the smoothness given by (6.115) is the objective function to be minimized.

[4]Readers can see Section 8.1 (Pages 240) of [70] for the derivation of the equation below and further discussion

Additionally the cluster assignment should be exclusive. Let $\boldsymbol{z}$ be one column of $\boldsymbol{Z}$. We can see that one product should be assigned to only one cluster, which can be represented by the following constraint[5]:

$$\boldsymbol{z}_k^\top \boldsymbol{z}_{k'} = 0 \quad \text{if} \quad k \neq k', \tag{6.117}$$

$$\boldsymbol{z}_k^\top \boldsymbol{z}_k = |C_k| \quad \text{otherwise}, \tag{6.118}$$

where $|C_k|$ is the number of products in the k-th cluster. This can be written in the following matrix form:

$$\boldsymbol{Z}^\top \boldsymbol{Z} = \begin{pmatrix} |C_1| & 0 & \cdots & 0 \\ 0 & |C_2| & 0 & \vdots \\ \vdots & 0 & \ddots & 0 \\ 0 & \cdots & 0 & |C_K| \end{pmatrix}. \tag{6.119}$$

We write the right hand side of (6.119) as $\boldsymbol{C}$:

$$\boldsymbol{Z}^\top \boldsymbol{Z} = \boldsymbol{C}, \tag{6.120}$$

Thus entirely the optimization problem can be formulated as follows:

$$\min_{\boldsymbol{Z}} \text{trace}(\boldsymbol{Z}^\top \boldsymbol{L} \boldsymbol{Z}) \quad \text{s.t.} \quad \boldsymbol{Z}^\top \boldsymbol{Z} = \boldsymbol{C} \tag{6.121}$$

Then we can write the following Lagrangian, relaxing the binary constraint of $\boldsymbol{Z}$:

$$L(\boldsymbol{Z}) = \text{trace}(\boldsymbol{Z}^\top \boldsymbol{L} \boldsymbol{Z}) - \lambda \text{trace}(\boldsymbol{Z}^\top \boldsymbol{Z} - \boldsymbol{C}) \tag{6.122}$$

By taking the derivative of the Lagrangian with respect to $\boldsymbol{Z}$ and setting that to zero, we can have the following eigenvalue problem:

$$\boldsymbol{L}\boldsymbol{Z} = \lambda \boldsymbol{Z} \tag{6.123}$$

Then by solving this eigenvalue problem, we can generate a 2D perceptual map by using the first two eigenvectors.

In summary, generating a perceptual map in this way can be summarized into the following steps:

Step 1. Compute $\boldsymbol{R}$ from input $\boldsymbol{X}$.

Step 2. Compute inner-product matrix $\boldsymbol{R}\boldsymbol{R}^\top$.

Step 3. Compute graph Laplacian $\boldsymbol{L}$ from (6.116).

Step 4. Run eigenvalue decomposition over $\boldsymbol{L}$.

Step 5. Generate a perceptual map by using the top two eigenvectors in **Step 4**.

[5]Readers can see Section 3.1 (Pages 15) of [70] for the derivation and discussion below

In **Step 2**, we can generate an adjacency matrix $\boldsymbol{R}\boldsymbol{R}^\top$. Instead of **Steps 1** and **2**, if the input is already a graph (i.e. adjacency matrix), say $\boldsymbol{W}$, and then tries to do partitioning (grouping) nodes, **Steps 3** and **4** can be done, and this is exactly the same as spectral clustering, which we introduced in Section 5.2.2. Then in spectral clustering, instead of **Step 5**, in which only the first two eigenvectors are selected, we run K-means clustering for partitioning nodes by using entire vectors. However, CA and spectral clustering share the main part of the above procedure, indicating that CA is clustering the products and visualizing the clustering result by using the first two eigenvectors only. Thus in other words, we can say that CA is equivalent to clustering while the main focus is on visualization.

Also we note that constrained K-means clustering over matrix $\boldsymbol{X}$ also solves eigenvalue decomposition to have clusters, and so CA is closely connected to constrained K-means clustering as well.

Further Modification of CA

Modification 1. Using kernel function for $\boldsymbol{R}$

As we have seen in **Interpretation 1**, CA uses the result (the first and second eigenvectors) obtained by solving the eigenvalue problem of $\boldsymbol{R}\boldsymbol{R}^\top$. Thus again solving CA is equivalent to solving the eigenvalue problem of $\boldsymbol{R}\boldsymbol{R}^\top$. $\boldsymbol{R}\boldsymbol{R}^\top$ is the inner-product matrix of rows (i.e. instances or products) of $\boldsymbol{R}$ (If we skip the preprocessing step of CA, $\boldsymbol{R}$ is $\boldsymbol{X}$ and $\boldsymbol{X}\boldsymbol{X}^\top$ is the inner-product matrix of the original input $\boldsymbol{X}$). This means that $\boldsymbol{R}\boldsymbol{R}^\top$ is the linear kernel of $\boldsymbol{R}$, implying $\boldsymbol{R}\boldsymbol{R}^\top$ can be replaced with a kernel function of $\boldsymbol{R}$.

A big advantage of using a kernel function is that we can incorporate background (prior) knowledge of the input, i.e. $\boldsymbol{R}$. For example, maybe we already know that some products are similar to each other, more than the similarity computed from their features, and this type of information can be incorporated to generate a kernel function over $\boldsymbol{R}$. If we do not have such a knowledge, a kernel function can be a simple similarity function computed from data, like the inner-product matrix $\boldsymbol{R}\boldsymbol{R}^\top$.

Then we can replace $\boldsymbol{R}\boldsymbol{R}^\top$ with kernel matrix $\boldsymbol{K}$ (a matrix of products $\times$ products), where (i,j)-element of $\boldsymbol{K}$ is $K(\boldsymbol{r}_i, \boldsymbol{r}_j)$ and $\boldsymbol{r}_i$ is the i-th row of $\boldsymbol{R}$. We then run eigenvalue decomposition over $\boldsymbol{K}$, and use the first and second eigenvectors, $\boldsymbol{u}_1$ and $\boldsymbol{u}_2$, respectively, as two axes to generate a perceptual map, as follows:

$$\boldsymbol{K}\boldsymbol{u}_1, \quad \text{and} \quad \boldsymbol{K}\boldsymbol{u}_2. \tag{6.124}$$

Also as we have done in KPCA, for new given instance $\boldsymbol{x}_{\text{new}}$, we generate kernel vector $\boldsymbol{k}_{\text{new}}$ and compute the following values to put this instance on the 2D space:

$$\boldsymbol{k}_{\text{new}}^\top\boldsymbol{u}_1, \quad \text{and} \quad \boldsymbol{k}_{\text{new}}^\top\boldsymbol{u}_2. \tag{6.125}$$

Procedure 6.15: PERCEPTUAL MAP BY KERNEL CORRESPONDENCE ANALYSIS.

Input: (customer $\times$ feature)-matrix $\boldsymbol{X}$
Output: Two axes and projected data

Compute standard residual matrix $\boldsymbol{R}$ from $\boldsymbol{X}$.;
Replace $\boldsymbol{R}\boldsymbol{R}^\top$ with kernel function matrix $\boldsymbol{K}$.;
Compute eigenvalue decomposition of $\boldsymbol{K}$.;
Use the first two columns of $\boldsymbol{U}$: $\boldsymbol{u}_1$ for $\boldsymbol{d}_1$ and $\boldsymbol{u}_2$ for $\boldsymbol{d}_2$.;
Project $\boldsymbol{X}$ by using the obtained two vectors into $(\boldsymbol{K}\boldsymbol{d}_1, \boldsymbol{K}\boldsymbol{d}_2)$.;

We can call this method *kernel correspondence analysis (KCA)*. **Procedure 6.15** shows a pseudocode, which summarizes the procedure of KCA.

The resultant perceptual map by the above procedure might be more realistic than just by using the similarities computed from given data or the perceptual map generated from given data, since a kernel function allows to take the advantage of both given data and prior knowledge.

Indeed a clear advantage of using a kernel function is to incorporate prior (background) knowledge into learning, while a drawback is scalability, since the kernel function must be totally filled without any missing information and then the kernel matrix cannot be huge. However on generating a perceptual map, this would not be a problem, since the number of products would be a moderate size, because if the number of products is so large, we will be unable to visualize all products.

Modification 2. Semisupervised learning: similarity constraint given. Another possible problem setting is that we have not only $\boldsymbol{X}$ but also some background information (or prior knowledge) on the similarity between instances partially. In other words, we plot products on the two dimensional space, using $\boldsymbol{X}$ as well as prior knowledge, which is supposed to be not complete like a kernel function but a *similarity matrix* of products $\times$ products.

We assume that the given similarity matrix is not perfect and only partial information on the similarity between products. Then we can regard this similarity matrix as an adjacency matrix $\boldsymbol{W}$ (or a graph).

On the other hand, the problem of CA is regarded as the eigenvalue problem of matrix $\boldsymbol{R}\boldsymbol{R}^\top$, meaning that this eigenvalue problem is derived from the following Lagrangian:

$$L(\boldsymbol{U}) = \boldsymbol{U}^\top\boldsymbol{R}\boldsymbol{R}^\top\boldsymbol{U} - \lambda\boldsymbol{U}^\top\boldsymbol{U} \quad (6.126)$$

and by setting the partial derivative of this Lagrangian with respect to $\boldsymbol{U}$ to zero, the eigenvalue problem (6.114) was obtained.

Procedure 6.16: PERCEPTUAL MAP BY SEMI-SUPERVISED CORRESPONDENCE ANALYSIS.

Input: (product $\times$ feature)-matrix $\boldsymbol{X}$; similarity matrix $\boldsymbol{W}$
Output: perceptual map

Preprocess data by transforming $\boldsymbol{X}$ into $\boldsymbol{R}$ by using (6.107).;
Compute graph Laplacian $\boldsymbol{L}$ from $\boldsymbol{W}$.;
Run eigenvalue decomposition over $(\boldsymbol{R}\boldsymbol{R}^\top + \lambda \boldsymbol{L})$ as in (6.129).;
Generate the 2D space by using the first two columns of $\boldsymbol{U}$, i.e. eigenvectors $\boldsymbol{u}_1$ and $\boldsymbol{u}_2$, as $\boldsymbol{d}_1$ and $\boldsymbol{d}_2$, respectively.;
Project products onto 2D space by $(\boldsymbol{R}\boldsymbol{R}^\top + \lambda \boldsymbol{L})\boldsymbol{d}_1$ and $(\boldsymbol{R}\boldsymbol{R}^\top + \lambda \boldsymbol{L})\boldsymbol{d}_2$.;

Then the prior knowledge on the product similarity can be added as a constraint to the above Lagrangian. That is, we can think about the smoothness of $\boldsymbol{U}$ over the adjacency matrix (or graph) $\boldsymbol{W}$ as follows:

$$\boldsymbol{U}^\top \boldsymbol{L} \boldsymbol{U}, \tag{6.127}$$

where graph Laplacian $\boldsymbol{L} = \boldsymbol{D} - \boldsymbol{W}$, $\boldsymbol{D}$ is a diagonal matrix and diagonal element $\boldsymbol{D}_{ii} = \sum_j \boldsymbol{W}_{ij}$.

Overall we can set up the Lagrangian as follows:

$$L(\boldsymbol{U}) \quad = \quad \boldsymbol{U}^\top \boldsymbol{R}\boldsymbol{R}^\top \boldsymbol{U} - \lambda_1 \boldsymbol{U}^\top \boldsymbol{U} + \lambda_2 \boldsymbol{U}^\top \boldsymbol{L} \boldsymbol{U}, \tag{6.128}$$

where both λ_1 and λ_2 are regularization coefficients (hyperparameters), and in particular λ_2 controls the effect by separately given adjacency matrix, more generally so-called *side information.*

Again by setting the partial derivative of this Lagrangian with respect to $\boldsymbol{U}$ to zero, we have the following eigenvalue problem:

$$(\boldsymbol{R}\boldsymbol{R}^\top + \lambda_2 \boldsymbol{L})\boldsymbol{U} = \lambda_1 \boldsymbol{U} \tag{6.129}$$

We can solve this eigenvalue problem and the first and second eigenvectors can be used for generating the two-dimensional space. Overall the procedure is as follows:

Step 1. Compute residuals $\boldsymbol{R}$ from input $\boldsymbol{X}$.

Step 2. Compute graph Laplacian $\boldsymbol{L}$ from given similarity matrix $\boldsymbol{W}$.

Step 3. Solve eigenvalue decomposition of $(\boldsymbol{R}\boldsymbol{R}^\top + \lambda_2 \boldsymbol{L})$.

Step 4. Generate a perceptual map by using the top two eigenvectors in **Step 3.**

Procedure 6.16 shows a pseudocode including the above four steps.

6.4.3 Finding the Suitable Place for a New Product against Competitors in Market

In Sections 6.3.1 we showed approaches to evaluate the quality of segments. In this evaluation, we used customer features given for training data or some additional feature to evaluate the segment. We may not have selected good segments properly under this situation of only limited data. On the other hand, however, in Section 6.4.2, we showed both competitors and also customer segment ideal points (CSIPs) in perceptual map, i.e. the space in which segments are placed,

Given a new product, we address the problem of selecting the segment or more specifically deciding the location of the new product, considering the competitors and also CSIPs. That is, there are two types of data: 1) one is competitors, which should be avoided and 2) the other is CSIPs, equivalent to customer preferences, showing the area where the customers exist. We then address to a problem of finding the location for a new product to be placed in market, considering the above two points. The problem setting is further clarified below.

Input:

1. (products $\times$ features)-matrix $\boldsymbol{X}$, where products are competitors. We have N competitors, $\boldsymbol{x}_1, \ldots, \boldsymbol{x}_N$, i.e. $\boldsymbol{X} = (\boldsymbol{x}_1, \ldots, \boldsymbol{x}_N)^\top$. Let M be the size of features. If we use 2D perceptual map, simply $M = 2$

2. (CSIPs $\times$ features)-matrix $\boldsymbol{\Phi}$, where CSIPs have the same features as competitors. In other words, both competitors and CSIPs share the same data space. We have K CSIPs, $\boldsymbol{\phi}_1, \ldots, \boldsymbol{\phi}_K$, i.e. $\boldsymbol{\Phi} = (\boldsymbol{\phi}_1, \ldots, \boldsymbol{\phi}_K)^\top$.

Output:

1. $\boldsymbol{z}$, which is the vector of the new product. $\boldsymbol{z}$ has the same set of features as competitors and CSIPs. In other words, $\boldsymbol{z}$ shares the same data space as competitors and CSIPs. Please note that the problem is NOT selecting segments but optimizing the location of the product, i.e. $\boldsymbol{z}$.

We can then consider the following two necessary conditions, to formulate the problem.

Situation 1. Do not fight against existing competitors: As mentioned above, in the data space (such as perceptual map) we can find competitors already. When we decide the best segment and even the location in the segment, primarily we need try avoiding the fight against any existing competitors. This means that we would like to avoid close locations to competitors as much as possible.

Situation 2. Need to stay in customers: We need customers definitely. This means the location of $\boldsymbol{z}$ needs to be close to at least one of CSIPs.

Situation 1:

We can first check the similarity between each of all competitors and the new product. Taking the simplest similarity, this similarity can be computed as $\boldsymbol{x}_i^\top \boldsymbol{z}$ for the i-th competitor. Then the similarity against all competitors can be given as the following vector:

$$\boldsymbol{X}\boldsymbol{z}, \tag{6.130}$$

This similarity does not have clear range, and so we add some certain value to make all element values positive, as follows:

$$\boldsymbol{X}\boldsymbol{z} + \kappa_1 \mathbf{1}, \tag{6.131}$$

where $\mathbf{1}$ is the vector with all elements of 1 and κ_1 is a constant, which is set to be large enough to make all elements of the vector in (6.131) positives. Then each element of (6.131) is a similarity, taking a positive value.

The idea of Situation 1 is the new product should avoid the already existing competitors, meaning that the similarity between the new product $\boldsymbol{z}$ and each of all competitors should be small. This means that the similarity given by (6.131) should be small, leading to the following optimization problem:

$$\text{minimize} \quad (\boldsymbol{X}\boldsymbol{z} + \kappa_1 \mathbf{1})^\top (\boldsymbol{X}\boldsymbol{z} + \kappa_1 \mathbf{1}) \tag{6.132}$$

Situation 2:

The idea of Situation 2 is that $\boldsymbol{z}$ should be kept close to at least one of CSIPs. Thus again we check the similarity between each of all CSIPs and the new product, as follows:

$$\boldsymbol{\Phi}\boldsymbol{z} \tag{6.133}$$

Again we can keep these similarities positives, as follows:

$$\boldsymbol{\Phi}\boldsymbol{z} + \kappa_2 \mathbf{1}, \tag{6.134}$$

where again constant κ_2 needs to be large enough to make the similarities all positives.

Different from the competitors, the new product does not have to think about all CSIPs. Instead the new product should be close to at least one CSIP. Thus we introduce weight parameter, $\boldsymbol{w}^\top = (w_1, \ldots, w_M)$ over CSIPs, showing to which CSIPs the new product is close. Then we take the inner product between $\boldsymbol{w}$ and the above vector of similarity, as follows:

$$\boldsymbol{w}^\top (\boldsymbol{\Phi}\boldsymbol{z} + \kappa_2 \mathbf{1}) \tag{6.135}$$

Then at least one element of $\boldsymbol{w}$ should be large.

Formulation:

Procedure 6.17: FINDING THE MOST PROPER LOCATION OF A NEW PRODUCT.

Input: (products × features)-matrix $\boldsymbol{X}$; (CSIPs × features)-matrix $\boldsymbol{\Phi}$, hyperparameters λ_1, λ_2, κ_1, κ_2

Output: New vector $\boldsymbol{z}$; weights over CSIP $\boldsymbol{w}$

```
1 Initialize parameters z and w.;
2 repeat
      /* Estimate z, fixing w.                                    */
3     Estimate ẑ by using (6.147).;
      /* Estimate w, fixing Z.                                    */
4     Estimate ŵ by using (6.143) and (6.144). ;
5 until convergence;
6 Output ẑ, the optimal location of the new vector.
```

Parameters $\boldsymbol{z}$ and $\boldsymbol{w}$ should be regularized. $\boldsymbol{z}$ can take the regular L^2 norm. On the other hand, for $\boldsymbol{w}$, if the new product is close to at least one CSIP, that is enough and then $\boldsymbol{w}$ can be sparse, and so we can use L^1 norm for $\boldsymbol{w}$, as follows:

$$||\boldsymbol{z}||^2 = \boldsymbol{z}^\top \boldsymbol{z} = \text{Constant} \tag{6.136}$$

$$||\boldsymbol{w}||^1 = \sum_i w_i = 1 \tag{6.137}$$

We use all above derived terms to formulate the problem as follows:

$$\min_{\boldsymbol{z},\boldsymbol{w}} \quad (\boldsymbol{X}\boldsymbol{z} + \kappa_1 \mathbf{1})^\top (\boldsymbol{X}\boldsymbol{z} + \kappa_1 \mathbf{1}) \tag{6.138}$$

$$\text{subject to} \quad \boldsymbol{w}^\top (\boldsymbol{\Phi}\boldsymbol{z} + \kappa_2 \mathbf{1}) < \text{Const.} ||\boldsymbol{z}||^2 = \text{Const.}, ||\boldsymbol{w}||^1 = 1 \tag{6.139}$$

We can then write the following Lagrangian:

$$L(\boldsymbol{z}, \boldsymbol{w}) = (\boldsymbol{X}\boldsymbol{z} + \kappa_1 \mathbf{1})^\top (\boldsymbol{X}\boldsymbol{z} + \kappa_1 \mathbf{1}) \tag{6.140}$$

$$- \lambda_1 \boldsymbol{w}^\top (\boldsymbol{\Phi}\boldsymbol{z} + \kappa_2 \mathbf{1}) + \lambda_2 (\boldsymbol{z}^\top \boldsymbol{z} - \text{Const.}) + \lambda_3 (\sum_m w_m - 1) \tag{6.141}$$

Optimization:

We have two types of parameters, $\boldsymbol{z}$ and $\boldsymbol{w}$, and we are unable to optimize these two parameters at the same time. When we fix $\boldsymbol{w}$, the problem becomes a convex problem with respect to $\boldsymbol{z}$, while $\boldsymbol{z}$ is fixed, the problem can be written as follows:

$$\max_{\boldsymbol{w}} \boldsymbol{w}^\top (\boldsymbol{\Phi}\boldsymbol{z} + \kappa_2 \mathbf{1}) \ \text{ subject to } \ \sum_i w_i = 1 \tag{6.142}$$

The condition for $\boldsymbol{w}$ is the sum of elements is 1 (i.e. L^1 norm like LASSO), and each element of vector $(\boldsymbol{\Phi z}+\kappa_2\mathbf{1})$ takes a positive value. Thus this maximization problem can be solved just by choosing the biggest element in vector $(\boldsymbol{\Phi z}+\kappa_2\mathbf{1})$ and gives 1 to the corresponding element of $\boldsymbol{w}$ and zero to the other elements. That is,

$$\hat{w}_{i'} = 1 \qquad \text{if} \qquad i' = \arg\max_i v_i, \tag{6.143}$$

$$\hat{w}_i = 0 \quad \text{otherwise}, \tag{6.144}$$

where v_i is the i-th element of vector $\boldsymbol{v} = (v_1, \ldots, v_M)^\top = \boldsymbol{\Phi z} + \kappa_2\mathbf{1}$

On the other hand, regarding the $\boldsymbol{z}$ side, setting the partial derivative of the Lagrangian with respect to $\boldsymbol{z}$ to zero, we have the following:

$$\frac{\partial L(\boldsymbol{z}, \boldsymbol{w})}{\partial \boldsymbol{z}} = \mathbf{0} \tag{6.145}$$

$$\Rightarrow$$

$$\boldsymbol{X}^\top \boldsymbol{X z} + (\kappa_1 + \kappa_2)\boldsymbol{X}^\top\mathbf{1} - \lambda_1\boldsymbol{\Phi}^\top\boldsymbol{w} + \lambda_2\boldsymbol{z} = \mathbf{0}. \tag{6.146}$$

$$\Rightarrow$$

$$\hat{\boldsymbol{z}} = (\boldsymbol{X}^\top\boldsymbol{X} + \lambda_2\boldsymbol{I})^{-1}(\lambda_1\boldsymbol{\Phi}^\top\boldsymbol{w} - (\kappa_1+\kappa_2)\boldsymbol{X}^\top\mathbf{1}), \tag{6.147}$$

where $\mathbf{0}$ is the vector with all elements of zero.

Overall the optimization procedure is to repeat the following two alternately:

1. estimating $\boldsymbol{z}$ by (6.147), fixing $\boldsymbol{w}$.

2. estimating $\boldsymbol{w}$ by (6.143) and (6.144), fixing $\boldsymbol{z}$.

Procedure 6.17 shows a pseudocode, which summarizes the above procedure, including the algorithm for estimating $\boldsymbol{z}$ and $\boldsymbol{w}$, alternately. This procedure provides $\boldsymbol{z}$, i.e. the location of a new product in the space by competitors and CSIPs.

Further Modification

1. **Fixed segments** The above problem addresses to an optimization problem against both competitors and also customers (segments). In more detail, the setting was to avoid all competitors and pick up at least one of the segments (customers). These two were optimized. Practically, segments might not be selected so easily, because the segments to be selected might be considered from a wide viewpoints, such as profitability, size (number of customers), etc.

   #### i. Manual selection

   Thus one simple approach is just to select a preferable segment manually beforehand, meaning that we choose one element of vector $\boldsymbol{w}$ beforehand. For example, if the s-th segment was chosen, $w_s = 1$ and $w_{i|i\neq s} = 0$, and then out of all K CSIPs, only the s-th CSIP, $\boldsymbol{\phi}_s$, is selected. Then in the above procedure, optimizing $\boldsymbol{w}$ is removed, and then $\boldsymbol{z}$ is estimated analytically by the following rule:

$$\hat{\boldsymbol{z}} = (\boldsymbol{X}^\top \boldsymbol{X} + \lambda_2 \boldsymbol{I})^{-1}(\lambda_1 \boldsymbol{\phi}_s - (\kappa_1 + \kappa_2)\boldsymbol{X}^\top \mathbf{1}), \tag{6.148}$$

ii. Estimated segment attractivity

We already showed the way to evaluate the segment, using any given feature, in Section 6.3.1. For example, this method computes the quality of segment C_k, as non-negative value Quality (C_k), as follows:

$$\text{Quality}(C_k) = \frac{\text{trace}(\boldsymbol{X}^\top \boldsymbol{V} \boldsymbol{W}^{(k)} \boldsymbol{X})}{\text{trace}(\boldsymbol{X}^\top \boldsymbol{W}^{(k)} \boldsymbol{X})}, \tag{6.149}$$

where $\boldsymbol{V}$ is a diagonal matrix with the (i, i)-th element of given feature value for the i-th customer, and $\boldsymbol{W}^{(k)}$ is a matrix given in Fig. 6.1 (a), where the elements in the grayed part are 1 and the other parts are zero.

We can then write

$$Q_k = \text{Quality}(C_k). \tag{6.150}$$

Then we generate a matrix of showing segment quality. That is, let $\boldsymbol{Q}$ be a diagonal matrix, where the (i, i)-element $\boldsymbol{Q}_{ii}$ is Q_k, as follows:

$$\boldsymbol{Q} = \begin{pmatrix} Q_1 & 0 & \cdots & 0 \\ 0 & Q_2 & 0 & \vdots \\ \vdots & 0 & \ddots & 0 \\ 0 & \cdots & 0 & Q_K \end{pmatrix}. \tag{6.151}$$

We can use $\boldsymbol{Q}$ for weighing segments, i.e. CSIPs $\boldsymbol{\Phi}$, in (6.147), resulting in the following:

$$\hat{\boldsymbol{z}} = (\boldsymbol{X}^\top \boldsymbol{X} + \lambda_2 \boldsymbol{I})^{-1}(\lambda_1 (\boldsymbol{Q}\boldsymbol{\Phi})^\top \boldsymbol{w} - (\kappa_1 + \kappa_2)\boldsymbol{X}^\top \mathbf{1}), \tag{6.152}$$

6.5 Summary of Machine Learning for Target Marketing

6.5.1 Data Space on Customers and Products

A key point of target marketing is that we can generate a data space which can be shared by both customers and products. In fact this space can be generated by a unique property of data we raised above, which has the following connections:

$$\text{products} \xLeftrightarrow{\text{Survey}} \text{customers} \xLeftrightarrow{\text{Features (bases)}} \text{segments}. \tag{6.153}$$

On this data space, we can generate segments of customers (Section 6.2), evaluate the quality of segments (Section 6.3.1) and also examine the optimum new place in the space which satisfies relationships between (customer) segments and products (Section 6.4.3). Also a typical example of this data space is perceptual map. We have described a variety of methods to generate the perceptual map (Section 6.4.2). Note that usually perceptual maps are two-dimensional (for visualization), while our approach allows to generate a space with any dimensions.

Finally we consider an interesting problem setting, in which we explore the most proper location of a new product in data spaces (say perceptual map) under the two constraints: the location of new products is 1) not close to those of existing products, and 2) close to at least one of the CSIPs, i.e. the location where customers exist.

Chapter 7

Machine Learning for Relationship Marketing

7.1 Customer Relationship Management

As mentioned in Chapter 4, customer relationship management (CRM) focuses more on retained customers more than new customers.

In this chapter, we consider the same data (matrix with customers for rows and their features for columns) as target marketing, while we have another matrix with customers for rows and products for columns.

7.1.1 Learning (Customers × Products)-Matrix

In target marketing, as summarized in Section 6.1, the data we have considered were always matrices with columns for features (and rows for customers (, products or segments)). On the other hand, in CRM, we can have another matrix with rows for customers and columns for products, and vice versa. Practically a (customers × products) matrix can be generated by using customer purchase data, i.e. the purchased items by customers, so-called *market baskets* of customers (see Section 2.1.2) or *transaction data*, where each instance is a transaction. Then the generated (customers × products)-matrix is generally called an *user-item* matrix (so customers and products are called *users* and *items*, respectively), which has been well analyzed in business applications of machine learning (or data mining).

A key feature of this matrix is that usually the size is very large in both dimensions (i.e. both users and items), while the number of the filled elements (purchased items by users) is not necessarily large (because even heavy users do not buy all items ever), indicating the high sparsity of the matrix. Because of this sparse nature of user-item matrices, matrix completion would be practically the most important issue for user-item matrices. In fact matrix completion is equivalent to predicting unknown items to be purchased by users, which is exactly *recommendation* of items for users. In fact the recommendation engine has been the most important, key software for E-commerce sites on internet as well as real retailers, in the past twenty years [113].

Recommendation of items for users is exactly a machine learning problem with the input of an user-item matrix. Machine learning problem settings for recommendation can be divided into two types: *collaborative filtering* and *content-based filtering*. Simply speaking (or in a narrow sense[1]), collaborative filtering just uses the input user-item matrix only, while content-based filtering uses more information on users and items. One example is movie rating by users, for which a user-item matrix can be a matrix of scores for movies rated by watched users, and collaborating filtering uses this movie rating matrix only, while content-based filtering uses not only this matrix but also the information of users and movies, e.g. user demographic data and movie contents, like movie genres, starring, etc. Also there is the other setting, which combines both collaborative and content-based filtering, being called *hybrid filtering*[2] [3].

In this section, i.e. Section 7.1.1, we focus on collaborative filtering and explain standard machine learning methods, which can be and have been used in collaborative filtering over user-item matrices, On the other hand, content information of content-based filtering can be regarded as side information, resulting in the input of multiple matrices, sharing their dimensions. Chapter 8 shows machine learning methods for database marketing, which has the input of multiple matrices, and so we then describe methods for content-based filtering in Chapter 8, particularly Section 8.1 and later.

In collaborative filtering, we can think that user-item matrices have no labels, meaning that machine learning methods for user-item matrices are unsupervised learning, such as clustering.

Notations

Let $\boldsymbol{X} = (\boldsymbol{x}_1, \ldots, \boldsymbol{x}_N)^\mathsf{T} = (\boldsymbol{f}_1, \ldots, \boldsymbol{f}_M)$ be a user-item matrix, where $\boldsymbol{x}_i$ is the i-th user, $\boldsymbol{f}_j$ is the j-th item vector, and $\boldsymbol{X}_{ij}$ shows a real value or binary, meaning how many times the i-th user purchased the j-th item or simply binary if the i-th user bought the j-th item already.

Detecting Correlated Combinations between Users and Items

Practically useful questions to be solved would be what type of combinations between items, between users or even between users and items exist. We explain possible approaches for realizing these purposes.

Clustering (constrained K-means clustering)

[1]As mentioned in Introduction, the idea of collaborative filtering is to find similar customers and use the past purchase history of the similar customers, for which the given user-item matrix can be used. However in a wider sense, collaborative filtering uses purchase records of any number of matrices, and in the sense of using the input of multiple matrices, collaborative filtering can be overlapped with content-based filtering. In this book, we use the term, collaborative filtering, for the case of only one input matrix, and the term, content-based filtering for the input of multiple matrices.

[2]In this book, we will not consider this setting.

Table 7.1: (top) Original data of market baskets have rows for transactions (baskets) and columns of products (items). Each transaction shows a user's purchase each time. (bottom) By using the market basket data at the top, we can count how many times each user purchased an item or simply binary data showing if each user purchased an item or not. This is called a *user-item* matrix.

		Item 1	Item 2	Item 3	Item 4	...
Transaction 1	User B					
Transaction 2	User A					
Transaction 3	User B					
Transaction 4	User C					
Transaction 5	User A					
	...					

⇓

	Item 1	Item 2	Item 3	Item 4	...
User A					
User B					
User C					
...					

For unsupervised data, the first approach would be clustering, e.g. constrained K-means clustering, which we have explained in Section 5.2.1 for machine learning on vectors and Section 6.2.1 for segmentation in target marketing. Once again we will briefly explain constrained K-means clustering: First let $\boldsymbol{Z}$ be a cluster assignment matrix, which introduced in Section 5.2.1 as a matrix with rows for instances and columns for features, showing the cluster to which each instance belongs. The objective of clustering is to summarize given users into a finite number of clusters, and this objective can be interpreted to minimize the distance from each instance to the closest cluster center. This minimization problem can be written, by using $\boldsymbol{Z}$ and the squared distance (or L^2 norm) for the distance between instance $\boldsymbol{x}_i$ and its cluster center $\boldsymbol{\mu}_k$, as follows:

$$\min_{\boldsymbol{Z},\boldsymbol{\mu}} \sum_{k=1}^{K}\sum_{i=1}^{N} \boldsymbol{Z}_{ik}||\boldsymbol{x}_i - \boldsymbol{\mu}_k||^2 = \min_{\boldsymbol{Z}} \sum_{k=1}^{K}\sum_{j=1}^{N}\sum_{i=1}^{N} \boldsymbol{Z}_{ik}\boldsymbol{E}_{ij}\boldsymbol{Z}_{jk} \quad (7.1)$$

$$= \min_{\boldsymbol{Z}} \operatorname{trace}(\boldsymbol{Z}^\top \boldsymbol{E}\boldsymbol{Z}), \quad (7.2)$$

where the element of the i-th row and the j-th column of $\boldsymbol{E}$, i.e. $\boldsymbol{E}_{ij}$, can be given as follows:

$$\boldsymbol{E}_{ij} = \frac{1}{2N}||\boldsymbol{x}_i - \boldsymbol{x}_j||^2. \quad (7.3)$$

Thinking about the constraint on $\boldsymbol{Z}$, which is $\boldsymbol{Z}^\top\boldsymbol{Z} = \boldsymbol{D}$ (where $\boldsymbol{D}$ is a $(N \times N)$-diagonal matrix), the problem can be formulated as the following minimization problem:

$$\min_{\boldsymbol{Z}} \quad \text{trace}(\boldsymbol{Z}^\top \boldsymbol{E} \boldsymbol{Z}) \text{ subject to } \boldsymbol{Z}^\top \boldsymbol{Z} = \boldsymbol{D}. \tag{7.4}$$

Then the binary constraint of $\boldsymbol{Z}$ is relaxed, and the Lagrangian of this optimization algorithm can be given as follows:

$$L(\boldsymbol{Z}, \lambda) = \text{trace}(\boldsymbol{Z}^\top \boldsymbol{E} \boldsymbol{Z}) - \lambda \text{ trace}(\boldsymbol{Z}^\top \boldsymbol{Z} - \boldsymbol{D}). \tag{7.5}$$

Taking the derivative of the Lagrangian with respect to $\boldsymbol{Z}$ and setting that to zero, we can have the following eigenvalue problem:

$$\boldsymbol{E}\boldsymbol{Z} = \lambda \boldsymbol{Z}. \tag{7.6}$$

After solving this eigenvalue problem, we can run the regular postprocessing procedure over $\hat{\boldsymbol{Z}}$ to have the finally estimated $\hat{\boldsymbol{Z}}$. The pseudocode of the procedure is given in **Procedure 5.3**. By using clustering, we can capture the combinations of users in each captured cluster.

Biclustering (constrained K-means clustering for both users and items) In the above, the combinations can be found between users but not between items. To find combination of items, we can run clustering (constrained K-means clustering) over items also of the same matrix $\boldsymbol{X}$. First let $\boldsymbol{V}$ be the binary cluster assignment matrix of the item side, where $\boldsymbol{V}$ also keeps the constraint:

$$\boldsymbol{V}^\top \boldsymbol{V} = \boldsymbol{D}, \tag{7.7}$$

where $\boldsymbol{D}$ is a diagonal matrix with the size of $M \times M$ (so this $\boldsymbol{D}$ can be different from $\boldsymbol{D}$ for $\boldsymbol{Z}$). Also let $\boldsymbol{F}$ be the matrix, where the (i, j) element satisfying the following:

$$\boldsymbol{F}_{ij} = \frac{1}{2M} ||\boldsymbol{f}_i - \boldsymbol{f}_j||^2. \tag{7.8}$$

Then clustering the item side can be formulated as follows:

$$\min_{\boldsymbol{V}} \quad \text{trace}(\boldsymbol{V}^\top \boldsymbol{F} \boldsymbol{V}) \text{ subject to } \boldsymbol{V}^\top \boldsymbol{V} = \boldsymbol{D}. \tag{7.9}$$

Thus similar to the user side, by relaxing the binary constraint of $\boldsymbol{V}$, we can consider the following Lagrangian:

$$L(\boldsymbol{V}, \lambda) = \text{trace}(\boldsymbol{V}^\top \boldsymbol{F} \boldsymbol{V}) - \lambda \text{ trace}(\boldsymbol{V}^\top \boldsymbol{V} - \boldsymbol{D}). \tag{7.10}$$

Then again by setting the derivative of the Lagrangian with respect to $\boldsymbol{V}$ to zero, we can have the following eigenvalue problem:

$$\boldsymbol{F}\boldsymbol{V} = \lambda \boldsymbol{V}. \tag{7.11}$$

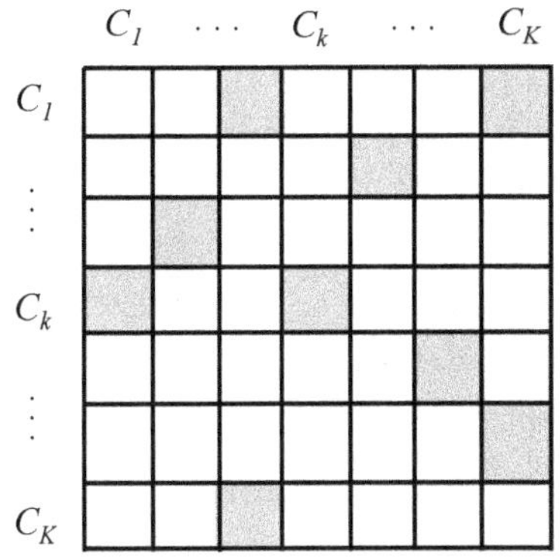

Figure 7.1: Schematic diagrams of biclustering, in which both the user and item sides are clustered by clustering (say, constrained K-means clustering).

We can estimate $\hat{\boldsymbol{V}}$ by solving this problem and also running the postprocessing algorithm given in **Procedure 5.2** over $\hat{\boldsymbol{V}}$, we can have the final cluster assignment $\boldsymbol{V}$. Then using $\boldsymbol{V}$, we can see the combination of items in each of captured item clusters.

We have shown clustering over each of the two sides, and then we note that these two clustering can run over an user-item matrix at the same time, and then we can have clusters for each side, independently. However the clusters of the user side and those of the item side are related with each other, because clustering users depends on the distance/similarity between user vectors, i.e. the similarity (in preferences) of purchased items between users. That is, clustering users depends on values by purchased items, and vice versa. Thus indeed we can run clustering over the two sides independently, and then we can see the resultant clusters, which can be specified by the both sides. Fig. 7.1 shows a schematic picture of the clusters obtained by running the clustering over both sides, and sort users and items, according to the obtained clusters, independently. In this figure, the gray parts are clusters (keeping some similar value, for example 1 if binary), while other white parts have only zero. As shown in the figure, clusters are kept for each of the two sides, and these clusters are in some cases, more detailed, such as two gray parts for one single row (or column). The point is that each of the gray parts can be specified by both sides, i.e. a group of users and a group of items. Then finally each gray part corresponds to the combination of users and items, which can be captured by clustering for the both sides. This clustering is called *biclustering*.

Procedure 7.1 shows a pseudocode of the above procedure of biclustering, where both the user and item sides of given user-item matrix $\boldsymbol{X}$ are clustered (Lines 1-2), then the both sides of $\boldsymbol{X}$ are sorted, according to clusters (Lines 3-4), and finally $\boldsymbol{X}$ is displayed on a 2D space (Line 5).

Clustering is useful for understanding the combinations of users, items and between users and items from observed data, explicitly. For biclustering, not only

Procedure 7.1: BICLUSTERING.

Input: User-item matrix $\boldsymbol{X}$

Output: User-side cluster assignment matrix $\boldsymbol{Z}$; item-side cluster assignment matrix $\boldsymbol{V}$

Run K`-means Clustering(`$\boldsymbol{X}$`)` to have estimated $\hat{\boldsymbol{Z}}$.;
Run K`-means Clustering(`$\boldsymbol{X}^\top$`)` to have $\hat{\boldsymbol{V}}$.;
Sort users in $\boldsymbol{X}$, according to $\hat{\boldsymbol{Z}}$.;
Sort items in $\boldsymbol{X}^\top$, according to $\hat{\boldsymbol{V}}$.;
Show sorted $\boldsymbol{X}$ in both of the two sides on 2D space.;

constrained K-means but also any type of clustering can be used. In fact clustering by probabilistic model, such as latent Dirichlet allocation, has been well used in the marketing literature [103, 102, 80]. The user-item matrix $\boldsymbol{X}$ is extremely sparse (again each user does never buy many items on user-item matrices), and if the data is too sparse, clustering would not be powerful enough for this data.

Matrix Completion: Estimating Missing Values in User-Item Matrix

Then a possible solution is we assume that a sparse user-item matrix (with many missing values) can be generated from a definitely fewer number of *factors* or *components*. We then explore methods for detecting factors from a given user-item matrix. This method is generally called *matrix factorization* or *matrix decomposition*.

Matrix Factorization

We can start with eigenvalue decomposition, which is one type of matrix factorization. Given matrix $\boldsymbol{X}$, eigenvalue decomposition can be written as follows:

$$\boldsymbol{X}\boldsymbol{u} = \lambda \boldsymbol{u}, \tag{7.12}$$

where $\boldsymbol{u}$ and λ are an eigenvector and an eigenvalue, respectively. By solving this problem, we can have multiple pairs of eigenvectors and eigenvalues $(\boldsymbol{u}_1, \lambda_1), \ldots, (\boldsymbol{u}_N, \lambda_N)$, in the descending order of eigenvalues, where eigenvalues have an orthonormal system, i.e. $\boldsymbol{u}_i^\top \boldsymbol{u}_j = 1$ if $i = j$; otherwise zero. Then $\boldsymbol{U} = (\boldsymbol{u}_1, \ldots, \boldsymbol{u}_N)$ and eigenvalue decomposition can be written as follows:

$$\boldsymbol{X}\boldsymbol{U} = \boldsymbol{U}\boldsymbol{\Lambda}, \tag{7.13}$$

where

$$\boldsymbol{\Lambda} = \begin{pmatrix} \lambda_1 & 0 & \cdots & 0 \\ 0 & \lambda_2 & 0 & \vdots \\ \vdots & 0 & \ddots & 0 \\ 0 & \cdots & 0 & \lambda_N \end{pmatrix}. \tag{7.14}$$

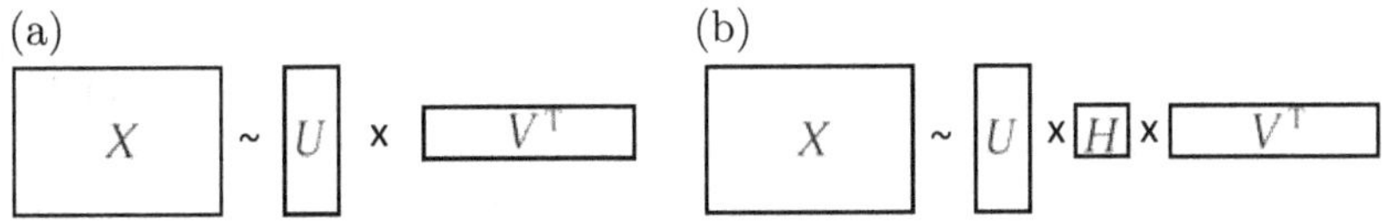

Figure 7.2: Factorizing matrix $\boldsymbol{X}$ into (a) two and (b) three low-rank matrices.

Eigenvectors $\boldsymbol{U}$ is an orthonormal system, and so $\boldsymbol{U}$ is a nonsingular (invertible) matrix, which means $\boldsymbol{U}$ has an inverse matrix. Thus from (7.13) we can write $\boldsymbol{X}$ as follows:

$$\boldsymbol{X} = \boldsymbol{U}\boldsymbol{\Lambda}\boldsymbol{U}^{-1}. \tag{7.15}$$

This is a special case of singular value decomposition, which is given as

$$\boldsymbol{X} = \boldsymbol{U}\boldsymbol{\Lambda}\boldsymbol{V}^{\top}. \tag{7.16}$$

These two decompositions are clearly one type of matrix decomposition. The assumption of the eigenvalue decomposition (or singular value decomposition) is that given $\boldsymbol{X}$ has the same rank as the size of $\boldsymbol{X}$, i.e. rank $= \min\{N, M\}$, which makes the above derivation possible. However, practically, given user-item matrix $\boldsymbol{X}$ has no guarantee on this point, i.e. rank $= \min\{N, M\}$. More importantly, a user-item matrix is sparse, having a large number of missing values. Thus it would be reasonable to think that the given matrix can be approximated by multiplying a few matrices with a lower rank than of given matrix $\boldsymbol{X}$. This idea has at least three nice properties: 1) even if a given matrix is huge, we can keep the given information within a limited space if we keep the rank of the low-rank matrices at some small value, meaning that we can save the space to record the original matrix, 2) we can think that the obtained low-rank matrices can extract the essential information from given data which are called *factors* in matrix factorization, and 3) so we first estimate the low-rank matrices so that they can reproduce given $\boldsymbol{X}$ as an approximation, and missing values in the given $\boldsymbol{X}$ can be also estimated by using the estimated low-rank matrices. The first and last parts of 3) correspond to unsupervised learning and prediction, respectively.

Fig. 7.2 shows schematic pictures of matrix factorization, where given $\boldsymbol{X}$ is approximated by (a) two or (b) three low-rank matrices which are called *bimatrix factorization* and *trimatrix factorization*, respectively.

Hereafter we focus on bimatrix factorization, which can be written as follows:

$$\boldsymbol{X} \sim \boldsymbol{U}\boldsymbol{V}^{\top}, \tag{7.17}$$

where $\boldsymbol{U}$ and $\boldsymbol{V}$ are low rank matrices with the size of $N \times K$ and $M \times K$, where the (i, j) element of $\boldsymbol{U}$ ($\boldsymbol{V}$) is $\boldsymbol{U}_{ij}$ ($\boldsymbol{V}_{ij}$). Importantly, K is significantly smaller than N and M: $K << N$ and $K << M$.

We now think about estimating $\boldsymbol{U}$ and $\boldsymbol{V}$ so that multiplying them approximates and reproduces given $\boldsymbol{X}$. If there are no constraints on $\boldsymbol{U}$ and $\boldsymbol{V}$, they can take any values, resulting in that a large number of possibilities for $\boldsymbol{U}$ and $\boldsymbol{V}$, which satisfy (7.17). Thus, using a real number p, we place the following constraints on the two low rank matrices $\boldsymbol{U}$ and $\boldsymbol{V}$:

$$||\boldsymbol{U}||^p = \text{Constant}, \qquad ||\boldsymbol{V}||^p = \text{Constant}. \tag{7.18}$$

On the other hand, the approximation of (7.17) can be the following minimization problem:

$$\min_{\boldsymbol{U},\boldsymbol{V}} ||\boldsymbol{X} - \boldsymbol{U}\boldsymbol{V}^\top||^p \tag{7.19}$$

We then put the above constraints and minimization problem together to formulate the problem of matrix factorization with $p = 2$ as a minimization problem as follows:

$$\min_{\boldsymbol{U},\boldsymbol{V}} ||\boldsymbol{X} - \boldsymbol{U}\boldsymbol{V}^\top||^2 \text{ subject to } ||\boldsymbol{U}||^2 = \text{Constant}, ||\boldsymbol{V}||^2 = \text{Constant}. \tag{7.20}$$

This is a biconvex problem of two parameters $\boldsymbol{U}$ and $\boldsymbol{V}$, for which we can repeat the following alternate procedure until convergence, to obtain a local optimum of the optimization problem of (7.20):

a) we estimate $\boldsymbol{U}$, fixing $\boldsymbol{V}$.

b) we estimate $\boldsymbol{V}$, fixing $\boldsymbol{U}$.

We can then consider the following Lagrangian (omitting Constant):

$$L(\boldsymbol{U},\boldsymbol{V}) = ||\boldsymbol{X} - \boldsymbol{U}\boldsymbol{V}^\top||^2 + \lambda(||\boldsymbol{U}||^2 + ||\boldsymbol{V}||^2). \tag{7.21}$$

Setting the partial derivative of the Lagrangian with respect to each of the two parameters to zero, we can have the following update rules for each of $\boldsymbol{U}$ and $\boldsymbol{V}$ (see Appendix of [70] for derivation) as follows:

$$\boldsymbol{U} \leftarrow \boldsymbol{X}\boldsymbol{V}(\boldsymbol{V}^\top\boldsymbol{V} + \lambda\boldsymbol{I})^{-1}, \tag{7.22}$$

$$\boldsymbol{V}^\top \leftarrow (\boldsymbol{U}^\top\boldsymbol{U} + \lambda\boldsymbol{I})^{-1}\boldsymbol{U}^\top\boldsymbol{X}. \tag{7.23}$$

Each element in user-item matrices is the purchased counts or binary value, showing the corresponding item is bought by the corresponding user. Thus all elements of $\boldsymbol{X}$ are nonnegative values. Matrix factorization with such a nonnegative matrix as the input and also nonnegative matrices as the output is called *nonnegative matrix factorization* (NMF). A standard manner for estimating parameters of low-rank matrices for NMF is so-called *multiplicative update rules*, which are given as follows (see Section 3.1.8 of [70] for more detailed explanation) for the (i,k)-elements of $\boldsymbol{U}$ and $\boldsymbol{V}$, i.e. $\boldsymbol{U}_{ik}$ and $\boldsymbol{V}_{ik}$, respectively:

$$\boldsymbol{U}_{ik} \leftarrow \boldsymbol{U}_{ik}\frac{(\boldsymbol{X}\boldsymbol{V})_{ik}}{(\boldsymbol{U}(\boldsymbol{V}^\top\boldsymbol{V} + \lambda\boldsymbol{I}))_{ik}}, \tag{7.24}$$

$$\boldsymbol{V}_{jk} \leftarrow \boldsymbol{V}_{jk}\frac{(\boldsymbol{X}^\top\boldsymbol{U})_{jk}}{(\boldsymbol{V}(\boldsymbol{U}^\top\boldsymbol{U} + \lambda\boldsymbol{I}))_{jk}}, \tag{7.25}$$

Procedure 7.2: NONNEGATIVE MATRIX FACTORIZATION.

Input: User-item matrix: $\boldsymbol{X}$; low-rank: K
Output: Low-rank matrices: $\boldsymbol{U}$, $\boldsymbol{V}$

```
Initialization: Initialize U and V.;
repeat
    Update U by (7.24).;
    Update V by (7.25).;
until convergence;
Output finally estimated Û and V̂.;
```

where $\boldsymbol{A}_{ij}$ indicates the (i,j)-th element of matrix $\boldsymbol{A}$.

Alternatively without considering the regularization term (i.e. $\lambda = 0$) in (7.22) and (7.22), the following update rules can be used as well:

$$\boldsymbol{U}_{ik} \leftarrow \boldsymbol{U}_{ik} \frac{(\boldsymbol{X}\boldsymbol{V})_{ik}}{(\boldsymbol{U}\boldsymbol{V}^{\mathsf{T}}\boldsymbol{V})_{ik}}, \tag{7.26}$$

$$\boldsymbol{V}_{jk} \leftarrow \boldsymbol{V}_{jk} \frac{(\boldsymbol{X}^{\mathsf{T}}\boldsymbol{U})_{jk}}{(\boldsymbol{V}\boldsymbol{U}^{\mathsf{T}}\boldsymbol{U})_{jk}}, \tag{7.27}$$

Procedure 7.2 shows a pseudocode of the procedure of running the above multiplicative update rules to estimate low-rank matrices, $\boldsymbol{U}$ and $\boldsymbol{V}$, of nonnegative matrix factorization with the input of $\boldsymbol{X}$.

On the other hand, the update rules given by (7.22) and (7.23) are called *additive update rules.* In general multiplicative update rules can be converged faster than additive update rules for the same optimization problem, because the change of parameters can be larger by multiplicative update rules than additive update rules.

Again matrix factorization allows to approximate a given user-item matrix by low-rank matrices which can capture factors of the given matrix and also the given matrix can be reproduced by the estimated low-rank matrices. Also for example, $\boldsymbol{U}$ is a $(N \times K)$-matrix, in which each row is a user and each column is a factor vector. We can think each factor vector is an important vector to generate the given user-item matrix. Then for one factor vector, if there are one or more elements with absolutely large values, we can think that the users corresponding to these elements share some common property, captured by this factor. This would lead to knowledge discovery of detecting similar users who share the preference on purchasing items.

In addition, nonnegative matrix factorization keeps all elements of low rank matrices nonnegative, which is useful for knowledge discovery more, since this situation clarify the importance of elements more, like that values closer to zero are less important and larger values are more important.

Notes

1. **Discussion on interaction terms** Matrix factorization is regarded as being equivalent to a linear model, for which in statistics, so-called interaction terms, which consider the interaction between two or more given variables, are used to make the model non-linear. This model is thought to be useful in the sense of being easy to understand the effective interaction terms, while problems of this straightforward extension are pointed out for a type of user-item matrices [26].

7.1.2 Detecting Profitable Customers

RFM Analysis

In Chapter 4, for finding profitable customers, we introduced RFM analysis, which considers three dimensions: recency, frequency and monetary. These features eventually build a three-dimensional (3D) space (see Fig. 4.2), which is manually segmented and each segment was examined manually (see Section 4.1.2). We can think that these three dimensions are just three customer features, resulting in a (customers × features)-matrix $\boldsymbol{X}$, which however has only three features. Below, with the data of RFM analysis, we explore machine learning-based approaches for finding profitable customers, also considering the possibility of improving and/or generalizing these approaches by using more general (customers × features) matrices.

Unsupervised and supervised learning on RFM

Again the RFM analysis generates the 3D space by three dimensions, recency, frequency and monetary, and customers can be distributed in the 3D space. As we explained in Section 4.1.2, in a manual manner, each dimension is, for example, evenly divided into five segments, resulting in totally 125 (=5 × 5 × 5) segments, for which analysis can be done.

One possible machine learning approach is unsupervised learning for generating the segmentation which is totally the same as using clustering (for example, constrained K-means clustering) for target segmentation in target marketing, because this data space is a customer space. By doing so, we can generate more adaptive clusters (segments) against the distribution of customers on the 3D space than any manual segmentation. In fact the difference from market segmentation is the RFM analysis just confines only the three features, which are manually selected to be most useful.

Also if we have labels on customers, we can set a supervised learning problem, i.e. classification, of predicting profitable customers by using three features. We can use supervised learning methods, such as linear regression, to the customer data of three features and can see which feature (and/or what combination of features) is most relevant to profitable customers. Also after applying supervised learning (classification), we can have a trained classifier (or regressor) to predict the profitability of an arbitrary customer by using the trained classification (or regression) rules. Fig. 7.3 shows schematic pictures of (a) unsupervised and (b) supervised (classification, i.e. assuming that the label is the binary, showing if

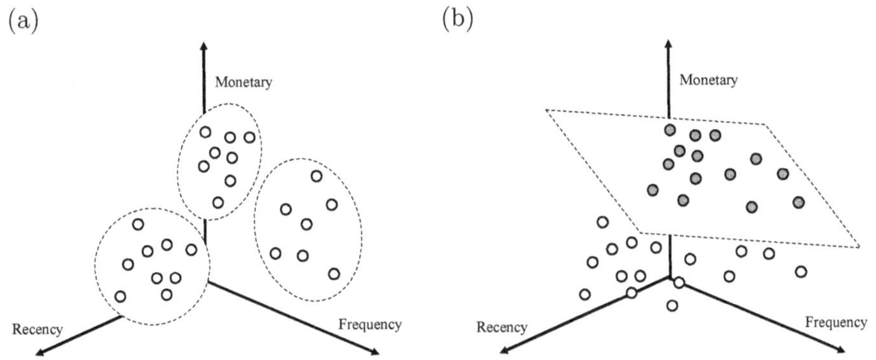

Figure 7.3: The data for RFM analysis can be regarded as customers with three features. Then depending on whether we can have labels over customers or not, we can apply (a) unsupervised and (b) supervised approaches to this data to detect profitable customers.

each customer is profitable or not,) settings over customers distributed on a 3D space, being generated by recency, frequency and monetary.

Profitable customers might be sometimes not like a binary label but multiple classes with an order or being continuous, for which we can use regression instead of classification.

Results of supervised as well as unsupervised learning would be useful, especially for a visual purpose, as shown in Fig. 7.3. In fact this is possible, because the number of features is only three, and these features are manually selected to be related with profitability well.

Generalization

In machine learning, using only three features is very primitive, and so we can consider approaches of extending the situation of only three features to a more general set of features and machine learning approaches for those features.

1. **Feature independence:** The three features: recency, frequency and monetary, are rather related (or correlated) with each other. For example, intuitively we can expect that if F (frequency) is high, R (recency) would be rather short and M (monetary) might be high as well. Obviously this shows that three features of the RFM analysis are not necessarily independent and cannot be equally considered (the result might be biased).

2. **More feature candidates:** These three features are selected manually, and they might be not independent as pointed out in the above. Thus it would be unsure if these three features are the best for unsupervised or supervised setting of the RFM analysis. A possible alternative strategy is to explore better features, for which we first prepare many features of customers and

run feature learning (See Section 5.1.2) over the features to select most influential (for profitability) three features. Also for example, in supervised learning, we can run linear regression directly over the dataset with many features, and the largest coefficients (over features) of linear regression can be selected as the most selective features. As such, it would be a good approach of using some machine learning models which can estimate parameters and at the same time select features out of given features.

3. **Time-dependent learning, thinking about long-term loyalty:** Detecting profitable customers must be time-dependent, in the sense that customers, who are just regular customers so far, may become highly profitable in the future. In fact the marketing funnel or customer journey mapping, introduced in Section 4.2.1, pointed out the customer change, and so the time-dependent property is already well recognized in existing marketing. In particular, detecting profitable customers should be discussed from viewpoints of long-term customer relationships. However RFM analysis lacks this point. Thus time-dependent features can be incorporated into data. We discuss machine learning strategies for time-series customer behaviors in the framework of marketing funnel in Section 7.2.

7.1.3 Customer Loyalty Satisfaction and Customer Lifetime Value

In order to understand the level of loyalty or satisfaction of customers, we need to have labels showing customer loyalty or customer satisfaction. One intuitive way is to use the answers of customer survey on some product (company, brand or service whatever) to assign labels to customers. However conducting a survey is laborious and also needs an unignorable amount of cost and time. Another drawback is that survey answers might not necessarily be correct customer responses.

However, there is a way we can have labels rather easily and then examine the customer satisfaction from data. A well used approach of this direction is *customer churn analysis* [60, 7].

Customer Churn Analysis

Churn analysis is the most prioritized problem in regular companies because of the strong connection to company profitability and also value [7]. Churn analysis focuses on past (and not current) customers of a membership site. A typical example can be found in telecommunication industry (where customers are mobile phone users), since acquiring customers is very competitive in this industry [2]. Churn analysis focuses on users who had a contract with some company and canceled the contract. Usually mobile phone companies keep customer information even after the customer canceled the contract, and so a company can use the current and past (not current) customer data, i.e. data of two classes of customers. These two classes can be a binary label for customers, and we can use supervised learning to generate a classifier which can classify a given arbitrary customer into two classes. Also if the classifier can provide features, which discriminate current customers from the past (not current) customers, these features would be key properties to retain the customers, more in detail, to improve customer satisfaction and boost customer loyalty.

Then once we generate customer data, i.e. a matrix of (customers $\times$ features) with binary labels for customers. We can run a training algorithm of any supervised learning (classification) over this data to generate a classifier (for example, [98] is a rather recent review of machine learning methods over the customer churn problem). Practically an important point is the classifier can be applied to any unknown customer vector to predict the customer class. In particular, we can apply the classifier to the current, existing customers to find prospective people who might leave in the near future [77]. Also by using a supervised learning model (classifier), which can provide classification rules explicitly, we can see one or more features and/or combinations of features which are correlated with the class label. As mentioned above, these features must be significant to improve customer satisfaction, boost customer loyalty and eventually retain the customers. This point would be connected to retention marketing to be described in Section 7.2.

Unique Property of Customer Churn Analysis: Class Imbalanceness

One unique property of customer churn analysis is the imbalanced data in terms of classes[3]. This is reasonable in nature, because customers who stopped the contract are only a small part of all customers, by which the so-called *churn rate*, i.e. the ratio of leaving customers to all customers, is a very small number, say 1-5% or less. This simply means the number of instances (customers) in one class is very limited and extremely small. This becomes a problem for regular supervised learning (classification), which assumes a good number of instances for each class.

Extreme class imbalanceness is a general issue in machine learning [42]. Below we introduce mainly three representative approaches to handle the imbalanced problem in the general supervised machine learning framework. They are typical and useful approaches, while note that they are not necessarily all but other approaches are possible.

Tackling Imbalanced Data

We show three approaches in supervised learning for the imbalanced data, in which we assume the ratio of instances in one class to all instances is extremely low, like 1%.

1. **Bootstrapping (+ ensemble learning)** Imbalanceness of labels causes a problem in supervised learning, and so it would be better to make the number of instances for each class balanced. To do so, a simple approach is to use *bootstrapping* or *sampling with replacement*, which for a given set,

[3] Also this issue can be shared by a lot of similar situations in other applications, for example: 1) insurance risk management and 2) fraud detection, where claims or frauds are very rare but the damage by them is very serious, and 3) diseased patients vs. healthy people, where serious disease would be very deadly.

Procedure 7.3: BOOTSTRAP + ENSEMBLE LEARNING FOR DATA IMBALANCENESS.

Input: Entire matrix labeled by major class A plus minor class B: $\boldsymbol{X} \leftarrow (\boldsymbol{X}_A + \boldsymbol{X}_B)$; K: #classifiers
Output: Classifier: $\hat{f}$

for $k \leftarrow 1$ **to** K **do** `/* down sampling */`
  $\hat{\boldsymbol{X}}_A \leftarrow \emptyset$.;
  repeat `/* generate a dataset by repeating bootstrapping */`
    Run Bootstrap($\boldsymbol{X}_A$) to generate new instance $\hat{\boldsymbol{x}}$.;
    Add $\hat{\boldsymbol{x}}$ to $\hat{\boldsymbol{X}}_A$.;
  until N_B *times*;
  Train a supervised learning algorithm f_k by $\boldsymbol{X} \leftarrow (\hat{\mathbf{X}}_A + \mathbf{X}_\mathbf{B})$.;
Generate the final classifier: $\hat{f}(\boldsymbol{x}) = \sum_k^K f_k(\boldsymbol{x})$ for arbitrary $\boldsymbol{x}$.

repeats randomly sampling an element out of the same originally given set. That is, we run bootstrapping, for example, the same number of times for each class so that the number of instances in each generated dataset (for each class) can be the same over all classes.

Regarding how many times we should run bootstrapping, we have two ways: *oversampling* and *undersampling*. That is, oversampling runs bootstrapping a larger number of times than the number of instances in the original dataset, while undersampling runs bootstrapping a fewer number of times than the original data size.

Bootstrapping does not guarantee that the dataset generated by bootstrapping has all instances in the original dataset (See Appendix A.1 of [70]). In other words, all information in the originally given data might not be put enough in the dataset generated by bootstrapping. Thus to extract the information from the original data good enough, we can repeat generating a training dataset (by bootstrapping). More simply, this idea leads to *ensemble learning*, which repeat running a supervised learning algorithm over one dataset multiple times.

In machine learning, two major ensemble learning algorithms are *bagging* [17] and *boosting* [89]. Interested readers in ensemble learning can refer Section 3.2 of [70].

Also both bagging and boostings are both applied to customer churn analysis (bagging [60, 83] and boosting [50]). More generally ensemble learning, particularly bagging was well used in data anlytics in marketing [39].

Overall a possible procedure is as follows:

1) First we repeat the following (Let A and B be major and minor classes, and N_B be the number of instances of a minor class):

i) run bootstrapping over instances in class A, N_B times, to generate N_B

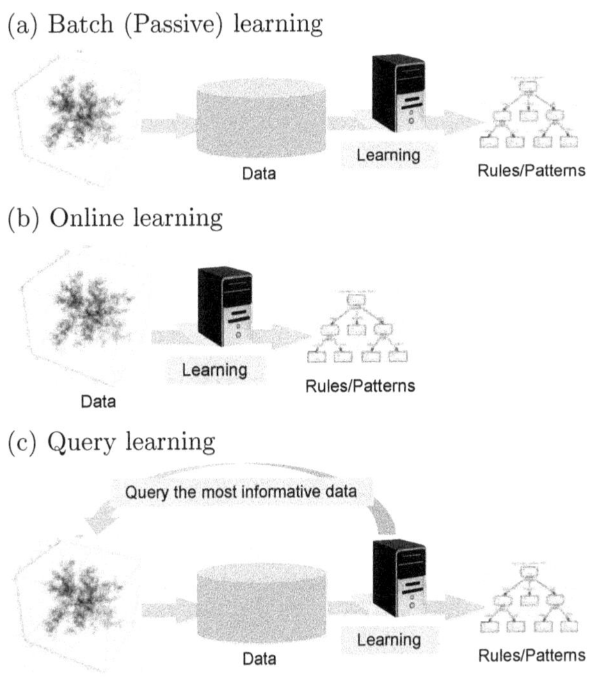

Figure 7.4: Three types of learning: (a) Batch (passive), (b) Online and (c) Query learning.

instances of class A (undersampling).

ii) train a supervised learning algorithm by using both the generated instances in class A and the original instances in class B.

2) We then generate a classifier by the majority vote of all classifiers generated by the above repetition.

Procedure 7.3 shows a pseudocode of the above procedure for data imbalancenss. In this pseudocode, we have two types of instances in this dataset: major and minor classes, by the number of instances. We let N_B be the number of instances in the minor class. Also function Bootstrap($\boldsymbol{X}$) randomly returns one instance of dataset $\boldsymbol{X}$.

2. Query learning The above procedure is based on random sampling. We may take a more strategic approach to sample instances. Below we consider this strategic manner of sampling.

A standard paradigm of machine learning is *passive learning*, in which a machine learning is just run over given data, and a learner is not allowed to use instances, which are not in the given data. On the other hand, the other paradigm is *active learning* or *query learning*, in which a learner can choose arbitrary instances, which are most necessary for learning the given data space as quickly as possible.

Procedure 7.4: QUERY LEARNING FOR DATA IMBALANCENESS.

Input: Entire matrix labeled by major class A plus minor class B: $\boldsymbol{X} \leftarrow (\boldsymbol{X}_A + \boldsymbol{X}_B)$; Q: #queries

Output: Classifier: $\hat{f}$

1 Generate initial data $\hat{\boldsymbol{X}}$ by bootstrapping for each class.;
2 **repeat**/* generate a dataset by repeating query learning */
3   Query_Learning($\boldsymbol{X}$) to generate new instance $\hat{\boldsymbol{x}}$.;
4   Add $\hat{\boldsymbol{x}}$ to $\hat{\boldsymbol{X}}$.;
5 **until** Q *times*;
6 Train a supervised learning method f by $\hat{\boldsymbol{X}}$,

More specifically, we can consider three types for learning manners. Fig. 7.4 shows these three types schematically. First, in (a), a regular manner is to obtain (raw) data and store them in a database, which is then used for learning. This is batch (passive) learning. Second, (b) is a special case, in which the raw data might be directly used for learning, skipping storing in a database. This is online learning. Third, (c) shows another special manner, in which a learner is permitted to request the instance which he/she needs the most. This allows the learning speed faster than the regular batch learning. This is called query or active learning.

Query learning is useful for the case in which obtaining a label is expensive and only a limited number of instances have labels. In marketing, for example, query learning is useful to explore the consumers' preferences for new products, where consumers and preferences are examples and labels, respectively [45]. Also query learning can be applied to detect marketing communities from large-scale networks, where larger networks become more sparse [22]. The idea of query learning is to select instances which are most *informative* to understand the data space. In other words, query learning raises instances most *uncertain*, as query instances, i.e. more plainly, the instances for which the label cannot be predicted easily by the current learner. For example, instances located at class boundaries in data space are likely to be new queries, and if their class labels are answered, the learner can understand the class boundary more in detail.

We borrow this idea of query learning of selecting instances to be added to the current data, although our setting is in the paradigm of passive learning and not active learning. Then instead of asking new instances, we can just select the instances closest to the queries.

Procedure 7.4 shows a pseudocode of the pseudocode of using query learning (in this pseudocode, function Query_Learning($\boldsymbol{X}$) returns one instance (as a query) out of input instance set $\boldsymbol{X}$) for sampling instances to generate one dataset, over which a supervised learning algorithm can be run. In this procedure, we first generate an initial dataset just by bootstrapping (Line 1). We then repeat running a query learning method to find the most in-

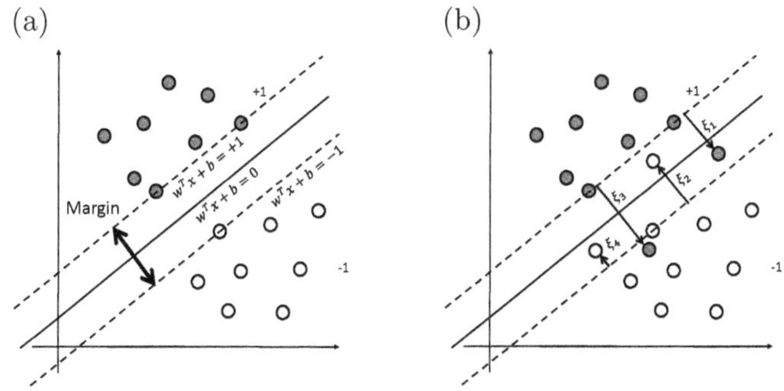

Figure 7.5: Maximizing the margin: (a) hard margin, (b) soft margin (From [70]).

formative instance, which should be added to the current dataset, a certain number of times (Lines 2-5). Finally a supervised learning method is run over the obtained dataset (Line 6).

This pseudocode can be modified to generate multiple datasets over which ensemble learning can be run, as in **Procedure 7.3**. Also note that this is one possible approach, and other ways of using query learning to the data imbalanceness issue would be also possible [71, 31, 8].

3. Label adaptive regularization: The problem is supervised learning, especially classification, for which we have a supervised learning model, such as linear regression, support vector machine (SVM), etc. These supervised learning models use regularizers for instances. One possible idea for the imbalanced data problem is to formulate one regularizer for each class. By doing this, we can control the regularization effect by each class, to relax the strong bias caused by imbalanceness. We will show just one example of this approach below.

For example, we introduced SVM in Sections 5.2.1 and 5.4.2. The SVM we explained was a version called *hard margin* (see Section 5.2.1). The hard margin version assumes instances of two classes can be separated completely, as shown in Fig. 5.3 (c). On the other hand, a practical situation is called *soft margin* (again see Section 5.2.1), in which instances cannot be separated clearly. Fig. 7.5 shows schematic pictures of (a) hard margin and (b) soft margin, when maximizing the margin in SVM.

For soft margin, instead of (5.23) and (5.24) in hard margin, we need to think about the following conditions for all instances:

$$\boldsymbol{\beta}^\mathsf{T}\boldsymbol{x} + b \geq +1 - \xi \qquad \text{for} \qquad y \geq 1. \tag{7.28}$$

$$\boldsymbol{\beta}^\mathsf{T}\boldsymbol{x} + b < -1 + \xi \qquad \text{for} \qquad y < -1. \tag{7.29}$$

where ξ_i $(i = 1, \ldots, N)$ can be defined for all instances with the possibility of reducing the margin, and are called *slack variables*. Also in the objective function, minimizing $\frac{1}{2}||\boldsymbol{w}||^2$ can be replaced with the following, by using the slack variables:

$$\frac{1}{2}||\boldsymbol{w}||^2 + C\sum_{i}^{N}\xi_i. \tag{7.30}$$

Then the formulation for hard margin, given in (5.26), can be changed into the following formulation for soft margin, as follows:

$$\min_{\boldsymbol{w}} \quad \frac{1}{2}||\boldsymbol{w}||^2 + C\sum_{i}\xi_i \tag{7.31}$$

$$\text{subject to} \quad \alpha_i y_i(\boldsymbol{w}^\top \boldsymbol{x}_i + b) \geq 1 - \xi_i \quad \text{for all} \quad i. \tag{7.32}$$

$$\xi_i \geq 0 \quad \text{for all} \quad i. \tag{7.33}$$

In this formulation of the optimization problem, the second term of (7.31) is summed over all instances, which include instances with both binary classes, say the positive and negative classes. One possible modification on this formulation, considering the imbalanceness problem is to formulate the regularizer of slack variables for every class, as follows:

$$\min_{\boldsymbol{w}} \quad \frac{1}{2}||\boldsymbol{w}||^2 + C_+\sum_{i|\boldsymbol{x}_i\in\mathcal{S}_+}\xi_i + C_-\sum_{i|\boldsymbol{x}_i\in\mathcal{S}_-}\xi_i, \tag{7.34}$$

where $\mathcal{S}_+$ and $\mathcal{S}_-$ are sets of instances with two classes, for example, positive and negative labels, respectively. Again by using this modification (which is called *Different Error Costs* (DEC) [106]), we can control the weight of regularizers depending on classes.

We note that other than the above three approaches, there are more heuristics based approaches for the class imbalance problem, for example, [21].

7.2 Retention Marketing: Learning Marketing Funnel

A key point of customer relationship management (CRM) is to focus on retained current customers more than attracting new customers, which makes retention marketing more important in CRM [7]. In particular the marketing funnel is a symbolic idea of retention marketing, and we can start with learning the marketing funnel from data. As shown in Figs. 4.3 and 4.4, a marketing funnel is a sequential transition of each customer. Models and methods under this assumption on customer acquisition are already well developed [20, 91].

The key assumption behind our model for retention marketing is that each customer will change his/her stage flexibly, not like the sequential property of general marketing funnel, but like a series of transitions over a graph, where each node is a stage in a general marketing funnel. We describe this assumption more carefully in the last part of Section 7.2.1, and then Section 7.2.2 shows a way of generating our data which can be state transitions over a graph of marketing funnel. Section 7.2.3 defines our problem setting, and Sections 7.2.4 and 7.2.5 present discrete and probabilistic approaches, respectively, for solving our problem.

7.2.1 Data Assumption

Figs. 4.3 and 4.4 showed examples of the marketing funnel. Both figures show a typical change of customer attitudes for/against a product (brand, company or service) by several stages (seven and six stages by Figs. 4.3 and 4.4, respectively). Each stage has its name (label), and by using the past customer behavior, customers can be assigned to one of these labels manually or through semi-manual analysis, such as RFM analysis (Sections 4.1.2 and 7.1.2). Thus we assume that all customers already are labeled by one of the around ten labels, like loyal customers. Then the objective of learning the marketing funnel is to understand the typical behavior of customers under the assumption that each customer has a label already, meaning that they are already categorized into one of the several stages at a time point.

The marketing funnel shows an ordered, stepwise stages, which would be a major variation of customers, while we can easily imagine that customers would not change along with this direction only, particularly one stage by one stage, always. Fig. 7.6 (a) is an example of the marketing funnel, taken from Fig. 4.4. For example, in Fig. 7.6 (a), a customer starting with the *Suspect* stage may jump into the *Customer* stage. Similarly a *Prospect* may jump into a *Client* or even a *Supporter*. Also from a viewpoint of retaining (holding or keeping) customers, some negative change of customers, for example, from a *Supporter* to a *Customer* or even a *Prospect*, would be also important to understand why this behavior happened.

In this regard, we do not think the marketing funnel is a simple one directional (left-to-right (Fig. 4.3) or bottom-to-top (Fig. 4.4)) model. Instead we need a more flexible marketing funnel, and then think the marketing funnel is a special graph, by considering the following points:

1) **Stage → node** Each stage in the marketing funnel (say Fig. 4.3) can be regarded as a unique *node* (or a *state*), in the sense that each node is different (with a different role).

2) **Any two stages can be connected** We assume that a customer can move from one stage to not only one but also any other stage[4]. Then any two nodes (originally stages) can be connected by a directed edge, including self-loops (meaning an edge which starts from a node and returns to the same node). By doing this, the original marketing funnel can be changed into a directed graph.

[4]This means a marketing funnel can be a complete graph. However of course depending on situations, we allow to use a non-complete graph. The simplest situation is "loyal user" would not become "Aware " in Fig. 4.3

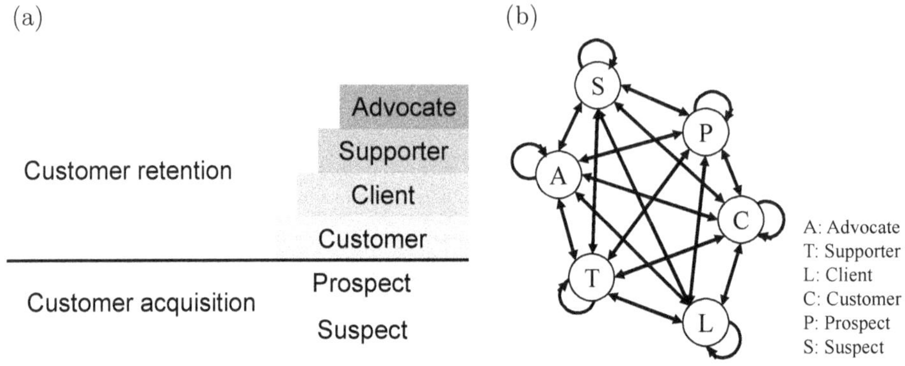

Figure 7.6: (a) An example of marketing funnel, taken from Fig. 4.4. (b) A directed, complete graph generated from the marketing funnel in (a), where one node corresponds to one of six stages in the marketing funnel and they are connected each other fully. This setting of fully-connected edges imply that customers can be moved from anywhere to anywhere.

That is, we model the marketing funnel as a *directed graph with unique nodes*, where nodes are labeled by stages. Fig. 7.6 (b) shows a directed graph, into which the marketing funnel shown in Fig 4.4 is converted. Below we discuss and explore possible machine learning problem settings over the directed graph with unique nodes and also approaches for solving the problems.

7.2.2 Data Acquisition

We assume that data is customer historical information on stages (labels) in the marketing funnel, meaning that each instance (customer) has a variable number of labels over timepoints, where the customer is labeled by a stage at each time point. A label corresponds to a stage, and each stage is a node in the given directed graph (with unique nodes) representing the marketing funnel. Thus we have two inputs:

Graph: One input is a directed graph with unique nodes, each corresponding to an unique label (stage).

String: The other input is, for each customer, a series of labels which shows a history of stages in the marketing funnel. This series of labels, i.e. a sequence of labels, results in a string. Thus, a set of sequences for multiple customers.

We call a move (of a customer) from a stage (node) to another stage (node) a *node transition*.

We explain these inputs further by using an example: first suppose that we have three stages (nodes): A, B and C. Then we assume that one customer (instance) changed his/her stages, for example, A $\rightarrow$ C $\rightarrow$ A $\rightarrow$ C. This means that

Table 7.2: (a) A sample dataset $\mathcal{D}$ of originally given strings (sequential customer behavior) and (b) $\boldsymbol{X}$ (generated from (a)) with description on the row (customers) and column (pairwise node (stage) transition) sides. The set of letters in strings is {A, B, C}.

(a)

String
ACAC
ABBC
ACBC
...

(b)

	A A	A B	A C	B A	B B	B C	C A	...
Customer1	0	0	2	0	0	0	1	...
Customer2	0	1	0	0	1	1	0	...
Customer3	0	0	1	0	0	1	0	...
...								

this customer was at stage A at the first time point and then moved to stage C at the second time point, and further moved to A and finally to C. We will write this ACAC for simplicity.

7.2.3 Problem Setting

Detecting Substring Patterns

Again one input is a directed graph with unique nodes, where this graph has M^2 edges for M nodes, if this graph is complete. Given an instance (corresponding to a customer), which is a string, for example, ACAC, which is equivalent to node transitions: A $\to$ C $\to$ A $\to$ C, on the given directed graph. Then if we have data on multiple customers with labels of stages, we have multiple strings, which can be originally given data which we write $\mathcal{D}$. Table 7.2 (a) shows a sample of dataset $\mathcal{D}$. Please note that in (a), all three strings have the same number, i.e. 4, of letters, while we can consider strings with an arbitrary length, like 3, 6 and 8, depending on how long the historical information on each customer has been recorded. That is, generally speaking, the methods we will introduce below can deal with strings with variable lengths.

Then let us consider a problem setting to detect substring patterns which are frequent in given data $\mathcal{D}$. A straightforward approach is to find frequent substrings by counting/enumerating the number of appearances in $\mathcal{D}$ for each of all possible substrings. This problem is called *frequent subsequence mining* (see Section 5 in [70]), which is a well investigated problem in data mining. Before explaining this problem, we define the *support* of a sequence as follows:

Definition 7.1 (support). *Given a set of sequences $\mathcal{D}$ and arbitrary sequence s, the number of sequences in $\mathcal{D}$ which have s is called the support of s against $\mathcal{D}$.*

Then we can define frequent subsequence mining as follows:

Definition 7.2 (frequent subsequence mining). *Given set of sequences $\mathcal{D}$, enumerate all (continuous) subsequences with the support against $\mathcal{D}$, larger than a certain prefixed amount, which is called minimum support or min_sup.*

Procedure 7.5: GRAPH-BASED LEARNING MARKETING FUNNEL: DISCRETE APPROACH.

Input: (instance × edge)-Matrix: $\boldsymbol{X}$; #patterns generated: K
Output: K substring patterns

```
Compute X^T X.;
Estimate K eigenvectors by solving the eigenvalue problem, given in (7.44).;
Select top K eigenvectors: ŵ^(1), ..., ŵ^(K).;
foreach ŵ_k do /* examine node transitions in a pattern */
    Check larger coefficients and corresponding node transitions, which
    should be in the k-th substring (node transitions) pattern.;
```

Frequent subsequence mining allows to output all subsequences frequent in a given set of sequences. There are however two issues on this approach:

Computationally high burden: We need to traverse a huge number of subsequences, which requests high computational cost always.

No summary: Frequent subsequence mining gives us frequent sequences only, which cannot be summarized at all. Instead it would be better to have summarized results on frequently appearing sequences.

Thus we now think about more efficient, machine learning-driven methods for inferring frequent substring patterns in given data more easily.

7.2.4 Discrete Approach

Eigenvalue Decomposition

From originally given dataset $\mathcal{D}$ of strings, we first generate matrix $\boldsymbol{X}$, in which each row is one instance, and each column corresponds to a node transition (directed node pair), like AC. That is, the number of columns is the number of edges, i.e. M^2, where M is the number of all nodes, and if an instance is, for example, ACAC, the columns, corresponding to AC and CA, take 2 and 1, respectively, and all other columns are zero. Suppose that the number of rows (instances) is N, $\boldsymbol{X}$ is a $(N \times M^2)$ matrix. Fig. 7.2 (b) shows an example of $\boldsymbol{X}$, which is generated from original string dataset $\mathcal{D}$, which is shown in (a).

Thus $\boldsymbol{X}$ shows the number of node transitions in each given sequence, and we call this matrix a *node transition matrix*. Let $\boldsymbol{w}$ be a real-valued (parameter) vector with the size of M^2 for M nodes of the given graph, meaning that each element of $\boldsymbol{w}$ is the coefficient of each directed edge in a directed graph. Then $\boldsymbol{w}$ should be estimated from given data so that if a directed edge appears in given data so frequently, the element corresponding to this edge in $\boldsymbol{w}$ should be larger. In other words, more generally, $\boldsymbol{w}$ should be trained to make smooth over $\boldsymbol{X}$.

For each instance, i.e. row vector $\boldsymbol{x}$ of $\boldsymbol{X}$, we can compute *similarity* between this instance and the weight vector by cosine similarity $\boldsymbol{x}^\top \boldsymbol{w}$. Let $\boldsymbol{x}_i$ be the i-th instance, and then $\boldsymbol{X}$ can be written as follows:

$$\boldsymbol{X} = (\boldsymbol{x}_1, \ldots, \boldsymbol{x}_N)^{\mathsf{T}} \tag{7.35}$$

Then again we need to maximize each similarity between one of N instances and $\boldsymbol{w}$, for which we can think that the square of each similarity should be maximized. Also this is equivalent to maximizing the sum of each squared similarity over all instances, since each instance can be independently computed in this sum. Then this sum can be modified as follows:

$$\begin{aligned}
|\boldsymbol{x}_1^{\mathsf{T}}\boldsymbol{w}|^2 + \cdots + |\boldsymbol{x}_N^{\mathsf{T}}\boldsymbol{w}|^2 &= (\boldsymbol{x}_1^{\mathsf{T}}\boldsymbol{w})^{\mathsf{T}}\boldsymbol{x}_1^{\mathsf{T}}\boldsymbol{w} + \cdots + (\boldsymbol{x}_N^{\mathsf{T}}\boldsymbol{w})^{\mathsf{T}}\boldsymbol{x}_N^{\mathsf{T}}\boldsymbol{w} && (7.36)\\
&= \boldsymbol{w}^{\mathsf{T}}\boldsymbol{x}_1\boldsymbol{x}_1^{\mathsf{T}}\boldsymbol{w} + \cdots + \boldsymbol{w}^{\mathsf{T}}\boldsymbol{x}_N\boldsymbol{x}_N^{\mathsf{T}}\boldsymbol{w} && (7.37)\\
&= \boldsymbol{w}^{\mathsf{T}}\sum_i(\boldsymbol{x}_i\boldsymbol{x}_i^{\mathsf{T}})\boldsymbol{w} && (7.38)\\
&= \boldsymbol{w}^{\mathsf{T}}\boldsymbol{X}^{\mathsf{T}}\boldsymbol{X}\boldsymbol{w} && (7.39)\\
&= (\boldsymbol{X}\boldsymbol{w})^{\mathsf{T}}\boldsymbol{X}\boldsymbol{w} && (7.40)\\
&= ||\boldsymbol{X}\boldsymbol{w}||^2 && (7.41)
\end{aligned}$$

That is, maximizing $|\boldsymbol{x}_1^{\mathsf{T}}\boldsymbol{w}|^2+\cdots+|\boldsymbol{x}_N^{\mathsf{T}}\boldsymbol{w}|^2$ is equivalent to maximizing $\boldsymbol{X}\boldsymbol{w}$. Thus the objective is to maximize $\boldsymbol{X}\boldsymbol{w}$, practically $||\boldsymbol{X}\boldsymbol{w}||$, and $\boldsymbol{w}$ should be regularized. Thus we can formulate the optimization problem as follows:

$$\max_{\boldsymbol{w}} ||\boldsymbol{X}\boldsymbol{w}||^2 \text{ subject to } ||\boldsymbol{w}||^2 = \text{Constant}. \tag{7.42}$$

Using $||\boldsymbol{X}\boldsymbol{w}||^2 = (\boldsymbol{X}\boldsymbol{w})^{\mathsf{T}}\boldsymbol{X}\boldsymbol{w} = \boldsymbol{w}^{\mathsf{T}}\boldsymbol{X}^{\mathsf{T}}\boldsymbol{X}\boldsymbol{w}$, we can have the following Lagrangian to solve the optimization problem given by (7.42).

$$L(\boldsymbol{w}) = \boldsymbol{w}^{\mathsf{T}}\boldsymbol{X}^{\mathsf{T}}\boldsymbol{X}\boldsymbol{w} - \lambda(\boldsymbol{w}^{\mathsf{T}}\boldsymbol{w} - \text{Constant}), \tag{7.43}$$

where λ is a Lagrangian multiplier.

We set the gradient of Lagrangian $L(\boldsymbol{w})$ with respect to $\boldsymbol{w}$ equal to zero, and this results in an eigenvalue problem, as follows:

$$\frac{dL(\boldsymbol{w})}{d\boldsymbol{w}} = 0 \qquad \Rightarrow \qquad \boldsymbol{X}^{\mathsf{T}}\boldsymbol{X}\boldsymbol{w} = \lambda\boldsymbol{w}. \tag{7.44}$$

We can solve this eigenvalue problem and then each obtained eigenvector contains node transition patterns. Note that the k-th eigenvector $\hat{\boldsymbol{w}}^{(k)}$ has M^2 coefficients over M^2 node transitions (edges). We can use the obtained eigenvector as follows: larger coefficients in the k-th weight $\hat{\boldsymbol{w}}^{(k)}$ indicate the corresponding node transitions are part of the k-th substring pattern in the graph. We call this method with the above procedure, *Node Transition Matrix Decomposition* (NTMD).

Fig. 7.7 shows sample results obtained by showing edge weights $\hat{\boldsymbol{w}}_k (k = 1, 2, 3)$ over the graph shown in Fig. 7.6 (b). From this figure, we can see that (a) shows rather normal transitions from Prospect and Customer to Client, and (b) also

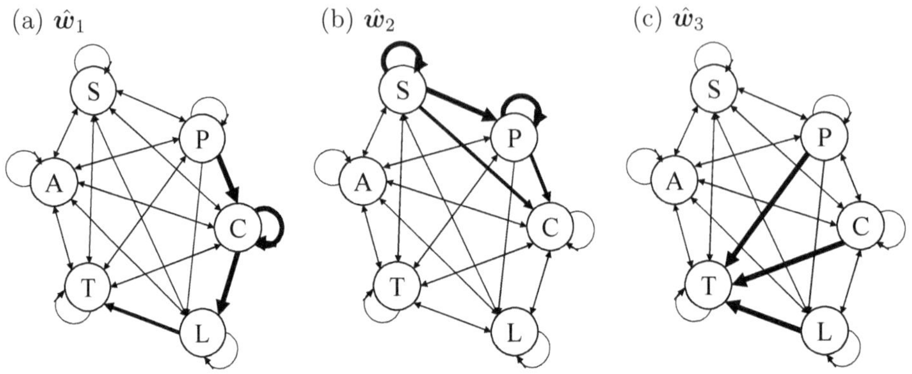

Figure 7.7: Schematic example pictures when we displayed the obtained edge weights (coefficients) (a) $\hat{\boldsymbol{w}}_1$, (b) $\hat{\boldsymbol{w}}_2$ and (c) $\hat{\boldsymbol{w}}_3$ over the graph shown in Fig. 7.6 (b).

shows transitions from Suspect to Prospect and Customer, while (c) shows transitions to Supporter which are not like standard transitions but more abrupt jumps from Prospect or Customer to Supporter. Thus, we can see that by optimizing edge weights $\boldsymbol{w}$, we can obtain frequent node transitions over the graph, by which we can see frequent subsequences in most cases. This result shown in Fig. 7.7 would be very useful for understanding customer behaviors over the stages in retention marketing. We note that this analysis can be done for any types of structures of the marketing funnel.

Notes

1. **Size of $\boldsymbol{X}^\top\boldsymbol{X}$: $\boldsymbol{X}$** is the matrix of instances × edges, where the number of edges reaches M^2 for M nodes in the given graph. Then the size of $\boldsymbol{X}^\top\boldsymbol{X}$ becomes $M^2 \times M^2$, which sounds very huge. However, in this application, nodes are stages in the original marketing funnel, and the number of stages is at most ten, meaning that M^2 is around 100 at maximum. This is a rather small size by which our discrete approach will never suffer from any computational or space complexity problem.

2. **Connection to Principal Component Analysis (PCA):** The above formulation and solution are totally the same as those of principal component analysis (PCA). For example, (7.42), (7.43) and (7.44) are the same as (5.85), (6.98) and (6.92), respectively.

   However, there are two different points:
   1) **Original motivation:** the idea of PCA is to project given data $\boldsymbol{X}$ into one dimensional (1D) space $\boldsymbol{Xw}$ so that the variance ($||\boldsymbol{Xw}||^2$) should be largest in the 1D space, meaning that the projected data should be as broad as possible in the 1D space. On the other hand, in this problem of estimating

frequent substring (node transition) patterns, we attempt to estimate weight vector $\boldsymbol{w}$ so that the similarity between each instance and the weight vector, i.e. $\boldsymbol{X}\boldsymbol{w}$, should be maximized. Thus although the formulation is the same, both started with different motivations.

2) Using estimated eigenvectors: after we obtained the eigenvectors, PCA usually uses only the top two of them to project data into the two-dimensional space defined by the two eigenvectors, while in the problem of estimating frequent node transitions, we check the weights/coefficients in each eigenvector and their values are visualized over the corresponding node transitions, resulting in principally K graphs, each corresponding to one eigenvector, for all K eigenvalues. We then can see multiple node transition patterns appearing in given data separately, even if they are overlapped.

7.2.5 Probabilistic Approach

Baseline Model

Each given instance is a string, such as ACAC, which is a series of node transitions. If we assign a probability to each node transition, this probability would be a conditional probability, such that letter C at time t conditionally depends upon letter A at time $t-1$. Then, for example, string $\boldsymbol{s}$ (= ACAC) can be schematically modeled as follows:

$$p(\boldsymbol{s}) = p(\mathrm{C}|\mathrm{A})p(\mathrm{A}|\mathrm{C})p(\mathrm{C}|\mathrm{A}). \tag{7.45}$$

In more general, we can define a probabilistic model of string $\boldsymbol{s} = s_1, \ldots, s_{|\boldsymbol{s}|}$ as follows:

$$p(\boldsymbol{s}) = p(s_1)\prod_i p(s_i|s_{i-1}) \tag{7.46}$$

In this model, due to the nature of conditional probabilities, the letter at time t depends upon the letter at time $t-1$ only. This property is called *1st-order Markov property*, and so this model is a *first-order Markov model*, hereafter *Markov model* for simplicity. The probability parameters of a Markov model are conditional probabilities, meaning that each variable depends upon another variable, where the point is that each variable depends upon only one variable, because of the 1st order Markov property.

To train a Markov model or estimate parameters of a Markov model, a simple approach is to estimate conditional probabilities directly from the data just by counting the number of each node transition. Let c_{AB} be the count of the transition from the node of letter A to the node of letter B over all given sequences. After obtaining c_{AB} for all possible A and B from given data, we can then compute the conditional probability from the node of letter A to the node of letter B, as follows:

$$p(\mathrm{C}|\mathrm{A}) = \frac{c_{\mathrm{AB}}}{\sum_{\mathrm{B}} c_{\mathrm{AB}}} \tag{7.47}$$

Note that these counts can be easily calculated from $\boldsymbol{X}$, e.g. from Fig. 7.2 (b).

Thus the procedure of this model can be summarized as follows:

Training

1) compute the count of a transition for all possible transitions.

2) compute the conditional probability, given in (7.47), by using the counts of transitions computed in 1).

Testing (prediction)

1) for new string $\boldsymbol{s}$ is given, compute the probability of $\boldsymbol{s}$ by using (7.46) and conditional probabilities corresponding to $\boldsymbol{s}$.

Finite Mixture Markov Model

One biggest problem of the above straightforward model is that the model is unable to capture multiple patterns separately, even if such patterns are in the given strings. In order to solve this problem, we add latent variable $\boldsymbol{z}$ to Markov model, which generates a *finite mixture Markov model* (F3M), where the structure can be given as follows:

$$p(\boldsymbol{s}) = \sum_{\boldsymbol{z}} p(\boldsymbol{z})p(\boldsymbol{s}|\boldsymbol{z}) \tag{7.48}$$

$$= \sum_{\boldsymbol{z}} p(\boldsymbol{z})p(s_1|\boldsymbol{z}) \prod_{i}^{|\boldsymbol{s}|} p(s_i|s_{i-1}, \boldsymbol{z}). \tag{7.49}$$

Analogically adding $\boldsymbol{z}$ has the same effect as doing clustering over given instances, i.e. strings.

Probability parameters can be estimated by, for example, the maximum likelihood criterion, for which a well-known approximation is the Expectation and Maximization (EM) algorithm.

EM algorithm for estimating parameters of this model, repeats the following two steps, until the convergence of parameter values.

E-step: Compute the posterior probability of latent variable $\boldsymbol{z}$:

$$p(\boldsymbol{z}|\boldsymbol{s}) = \frac{p(\boldsymbol{z})p(\boldsymbol{s}|\boldsymbol{z})}{\sum_{\boldsymbol{z}} p(\boldsymbol{z})p(\boldsymbol{s}|\boldsymbol{z})}. \tag{7.50}$$

M-step: Update parameters, $p(\boldsymbol{z})$, $p(s_i|\boldsymbol{z})$ and $p(s_i|s_j, \boldsymbol{z})$ by using posterior probabilities:

$$p(\boldsymbol{z}) \propto \sum_{\boldsymbol{s} \in \mathcal{D}} p(\boldsymbol{z}|\boldsymbol{s}), \tag{7.51}$$

$$p(s_i|\boldsymbol{z}) \propto \sum_{\boldsymbol{s} \in \mathcal{D}} \boldsymbol{I}_{\boldsymbol{s}, s_i} p(\boldsymbol{z}|\boldsymbol{s}), \tag{7.52}$$

$$p(s_i|s_j, \boldsymbol{z}) \propto \sum_{\boldsymbol{s} \in \mathcal{D}} \boldsymbol{X}_{\boldsymbol{s}, s_j s_i} p(\boldsymbol{z}|\boldsymbol{s}), \tag{7.53}$$

Procedure 7.6: GRAPH-BASED LEARNING MARKETING FUNNEL: PROBABILISTIC APPROACH: FINITE MIXTURE MARKOV MODEL.

Input: Given strings: $\mathcal{D}$; Markov model (graph): $\boldsymbol{G}$
Output: conditional probability parameters

```
Initialize parameters p(z), p(s_i|z) and p(s_i|s_j, z).;
repeat
    M-step: Update parameters by (7.51), (7.52) and (7.53).;
    E-step: Update posterior probabilities by (7.50).;
until convergence;
Output the estimated parameters.;
```

Table 7.3: Node Transition Matrix Decomposition (NTMD) and Finite Mixture Markov Model (F3M) both can generate multiple weighted graphs, as shown in Fig. 7.7. Also each graph of these multiple graphs has weighted edges, this table shows what each graph and weight correspond to, in the two methods, NTMD and F3M.

	NTMD	F3M
One graph	eigenvector	cluster
Edge weight	corresponding value of eigenvector	conditional probability
Priority	according to eigenvalues	all clusters equal

where $\mathcal{D}$ is the originally given set of strings, and $\boldsymbol{I}_{\boldsymbol{s},s_i}$ is given as follows:

$$\boldsymbol{I}_{\boldsymbol{s},s_i} \leftarrow 1 \quad \text{if} \quad s_i = s_1. \tag{7.54}$$

$$\boldsymbol{I}_{\boldsymbol{s},s_i} \leftarrow 0 \quad \text{otherwise.} \tag{7.55}$$

And also $\boldsymbol{X}_{\boldsymbol{s},s_j s_i}$ is the element in $\boldsymbol{X}$ which is specified by instance $\boldsymbol{s}$ for the row and the transition from s_j to s_i for the column. These two steps are easily derived from the analogy of *finite mixture model* (see Section 3 of [70]), the most typical probabilistic model for clustering.

Procedure 7.6 shows a pseudocode of the above EM algorithm of finite mixture Markov model (F3M).

After parameter estimation, for each value of latent variable $\boldsymbol{z}$, we can assign the values of estimated conditional probabilities over the edges of a given directed graph, which schematically leads to similar visualization to the three graphs shown in Fig. 7.6 (b).

Notes

1. **Size of given data** In both NTMD and F3M, the size of the given input, i.e. matrix, is automatically determined from the number of steps in the marketing funnel, also meaning that the number of nodes in the graphs in

Fig. 7.7 is kept as the same size as the number of steps in the marketing funnel.

2. **Difference of F3M from hidden Markov models (HMMs):**

   Hidden Markov models (HMMs) are a standard probabilistic model for various time-course data, particularly speech recognition [81]. Also HMMs have been used in marketing for, for example, time-varying marketing mix [53] and customer life time value measurement [84], etc. F3M has the same time complexity in estimating parameters as HMM. The difference of F3M from HMM is that HMM provides only one graph (generally called, *state transition model*, such as one graph in Fig. 7.7), while F3M provides multiple graphs as shown in Fig. 7.7). In other words, both are doing clustering: 1) each cluster is one graph in F3M, 2) clustering given sequences over one graph in HMM.

3. **Difference between NTMD and F3M:** Table 7.3 shows the summary of differences in the final graph visualization, between NTMD and F3M. As shown in the table, both models can generate multiple graphs, for which each graph corresponds to one eigenvector in NTMD and one cluster in F3M. Also each edge of one graph is weighted by the corresponding value of the eigenvector in NTMD and also the corresponding conditional probability of F3M. Finally multiple graphs (eigenvectors) by NTMD can be ordered according to eigenvalues, while clusters generated by F3M are all equal.

4. **Applications of F3M:** F3M have been applied to other domains also, such as web access pattern mining [18] and detecting active metabolic paths with gene expression in bioinformatics [72].

7.3 Market Communications

7.3.1 Optimizing Market Communication Channels

In Section 4.3, we showed at least eight different communication channels already known: 1) advertising, 2) sales promotion, 3) public relations and publicity, 4) events and experiences, 5) direct marketing, 6) interactive marketing, 7) word-of-mouth (WOM) marketing and 8) sales force.

First we need to consider two more things on market communication channels:

1. **Internet:** Currently internet, particularly social networks (or social media), would be the most powerful communication channels for marketers [44, 92, 15]. Thus we need to add one or more communication channels, resulting in totally nine or much more channels.

2. **Further detailed classification:** The above eight types are clearly major classification, and in each channel, there would be further details. For example, direct marketing has a large number of retailers, which are different

each other. Also internet has different types of channels like homepages, emails and social networks, etc.

Considering the original eight communication channels and the above two points, we can assume that we have a certain number (at least nine) of major channels, where each channel has a further number of detailed channels. Assuming that these communication channels can be clearly separated each other, we can consider several problem settings, in which the market communication channels are optimized so that the company should use the communication channels as such.

Supervised Setting

We can consider two practical applications, and both can be summarized into one supervised problem setting, in which we have some label ($\boldsymbol{y}$), like the total company profit, and then communication channels can be regarded as a set of variables to which values ($\boldsymbol{X}$) are assigned.

We can first consider a supervised setting, for which two different practical applications

1. **Company advertisement costs ($\boldsymbol{x}_i$) → profit (y_i):** The question is, for each company, what portfolio over the communication channels is the best to achieve a higher profit. More explicitly, the company would have the question on how much money should be spent on each communication channel within the limited budget.

   Each instance is a company, with a set of values, where each value is on how much money this company spent for each of all possible communication channels. Then after the promotion, the profit of the company can be measured. This information can be obtained by multiple companies (or the same company but at different time points). Then each company (or each time point) becomes one instance. That is, let $\boldsymbol{x}_i$ and y_i are the portfolio vector and the obtained profit of company i (or i-th time point), respectively. In this application, y_i is numerical, and so this is a regression problem.

2. **Customer approached by channels ($\boldsymbol{x}_i$) → purchase (y_i):** The above problem setting was a viewpoint from a company, and also a macroscopic viewpoint, in the sense that the profit is considered for not only on each customer but from the entire market. Then the second setting is microscopic, in the sense that the focus is on each customer. The question here is similar to the first setting. The question is what communication channels are most useful to make customers purchase some particular item.

   Each instance is a user (or customer), and the label (y_i) is if this user bought an item or not. Then each instance has a set of values over all communication channels, where each value shows how much time this user contacts the corresponding communication channel, resulting in $\boldsymbol{x}_i$. The label would be binary in this case, meaning that this is a classification problem.

Then in both cases, the problem is a straightforward supervised learning problem, with the following input

$$\boldsymbol{X} = (\boldsymbol{x}_1, \ldots, \boldsymbol{x}_N)^\top \tag{7.56}$$

$$\boldsymbol{y} = (y_1, \ldots, y_N)^\top \tag{7.57}$$

As a solution example of supervised learning, we explain linear regression (ridge regression: Section 5.2.1), where the coefficient parameter of this model is $\boldsymbol{\beta}$, i.e. weight vector over the communication channels. Then, following the description in Section 5.2.1, the objective function is to minimize the squared distance between $\boldsymbol{y}$ and $\boldsymbol{X}\boldsymbol{\beta}$:

Thus the problem can be formulated as follows:

$$\min_{\boldsymbol{\beta}} (\boldsymbol{y} - \boldsymbol{X}\boldsymbol{\beta})^\top (\boldsymbol{y} - \boldsymbol{X}\boldsymbol{\beta}) \text{ subject to } \boldsymbol{\beta}^\top \boldsymbol{\beta} = \text{Const.} \tag{7.58}$$

Solving this problem by using the method of Lagrange multipliers, we can have the analytical solution for updating $\boldsymbol{\beta}$ as follows:

$$\hat{\boldsymbol{\beta}} = (\boldsymbol{X}^\top \boldsymbol{X} + \lambda \boldsymbol{I})^{-1} \boldsymbol{X}^\top \boldsymbol{y} \tag{7.59}$$

From this result, we can see what communication channels should be used by using the estimated $\hat{\boldsymbol{\beta}}$. In fact we can use a variety of other supervised approaches instead. For example, just instead of squared regularization, we can use a more sparse regularization. Also more complex learning model, such as kernel ridge regression, support vector machine, etc. can be used. In fact the size of data would not be so large, by which a variety of supervised learning models can be applied.

Unsupervised Setting

Labels of supervised learning are sometimes too expensive to obtain, and then data have no label information (only $\boldsymbol{X}$ without $\boldsymbol{y}$). For example, in the above first application, a company would have data on how much cost this company spent for each communication channel, while this company may think that the profit might not be connected to the advertisement costs clearly. In this case the company might be interested more in analyzing given data ($\boldsymbol{X}$) only. Or the company may think the data should be analyzed first and then want to link the result to the labels.

In this case, the following three approaches would be possible:

1. Clustering: The most typical approach would be clustering, for example, constrained K-means clustering. As we have shown this approach well, given input matrix $\boldsymbol{X}$, we can use clustering assignment matrix $\boldsymbol{Z}$ (see Section 5.2.1) to formulate the problem:

$$\min_{\boldsymbol{Z}} \quad \text{trace}(\boldsymbol{Z}^\top \boldsymbol{E} \boldsymbol{Z}) \text{ subject to } \boldsymbol{Z}^\top \boldsymbol{Z} = \boldsymbol{D}, \tag{7.60}$$

where $\boldsymbol{E}$ and $\boldsymbol{D}$ are given in (5.40) and (5.43), respectively.

Solving this by the method of Lagrange multipliers, $\boldsymbol{Z}$ can be the solution of the following eigenvalue problem:

$$\boldsymbol{E}\boldsymbol{Z} = \lambda \boldsymbol{Z}. \tag{7.61}$$

From the cluster assignment, we can see group of users, which should be similar preferences each other.

Also we explained biclustering in Section 7.1.1 already, matrix $\boldsymbol{X}$ can be clustered not only by the customer side but also the communication channel side, where these two clustering can be performed independently, while biclustering would make each cluster more specific.

2. Matrix factorization: Matrix factorization was used to estimate missing values in given matrix. In general matrix factorization decomposes given matrix into low-rank matrices, with factors, from which we can reproduce the given matrix. That is, from the factors we can see the essential part of the given matrix.

Regardless of missing values in given $\boldsymbol{X}$, matrix factorization would be useful to know factors in low-rank matrices. We will here skip the procedure to estimate values in low-rank matrices, generated by matrix factorization (see Section 7.1.1).

3. Data integrative learning

For example, if instances are users, we can use a (users $\times$ items (products))-matrix further. That is, we can have two matrices, one being a (users $\times$ communication channels)-matrix and the other being a (users $\times$ items)-matrix. Then we can consider problem settings when we use these two matrices together. We will explain this type of data integrative learning in the next chapter in more detail.

Chapter 8

Machine Learning for Database Marketing: Data Integrative Approaches

Currently a wide variety of data are available and the entire amount of data is drastically increasing in many applications. This situation is true in marketing as well. In particular, this situation can be applied to relationship marketing, where various relationships, such as user-item and user-product, etc. can be considered.

Among different types of data, a typical data type is that one instance is a vector, showing a set of features of this instance, and in fact (users $\times$ products)-matrix in target marketing and (users $\times$ items)-matrix in relationship marketing are both data of vectors. Machine learning methods for vectors are well matured, meaning that a variety of methods exist for both supervised and unsupervised learning. We have showed them already in Section 5.2.1, focusing on basic approaches in machine learning.

Other than vectors, there exist various types of data, which can be converted into vectors. Importantly, however, converting other types of data into vectors is not always a good strategy, since some information might be lost during the conversion of another type of data into vectors. A typical example is a node in a graph, where nodes are instances and a set of instances is a graph. A graph is an adjacency matrix (Fig. 5.4), showing the connection (edges) between nodes. This is a square, symmetric matrix, in which the number of features reaches the number of nodes (instances), if each instance (row) of this matrix is considered to be a vector of features. When we use this matrix, if we think the length of each vector is too large and should be reduced, meaning that features are reduced, by which the original information might not be kept. In this sense, the conversion from another type of data into vectors might loose some information. A real-world example of nodes in a graph is social networks, in which users are nodes and a social network is a graph generated by connections among users. In fact this is a practical problem of combining different types of data on consumer behavior,

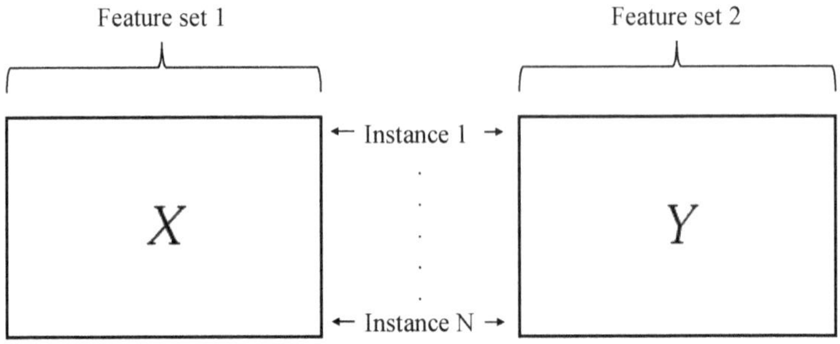

Figure 8.1: Multiview learning with two views: schematic diagram of multiview learning with two views, which share the same set of instances, while two views have different sets of features.

such as text, images, audio and video [67]. We then already explained machine learning methods for this type of data, i.e. nodes in a graph, in Section 5.2.2.

These two types of data are typical in marketing and can be given in marketing at the same time. For example, demographic data of users and also social networks can be both given, where each user is a vector of demographic data and at the same time a node in a graph, which corresponds to a social network. We then, in Section 5.2.3, present both supervised and unsupervised methods for this setting of each instance, being represented by a vector and a node in a network.

We can often meet the situation of having multiple data sets at the same time, and this situation is not necessarily the above case of having two typical datasets: one is a vector and the other is a node in a network. For example, each instance may have two different vectors, or the same set of features can be given to two different sets of instances. More generally we can have two matrices, which shares one dimension, which might be a set of instances or a set of features. Also the situation can be more developed into more than two matrices.

A typical example is again multiple matrices always with users for instances (columns) and features (rows) are different. For examples, features are demographic data or a set of purchased items, etc. This is an example of content-based filtering, in which not only an user-item matrix but also an matrix of user features and/or item contents can be given.

Another example was in Section 6.4.3, where input features are shared by two different matrices:

(product $\times$ feature)-matrix: $\boldsymbol{X}$

(CSIP $\times$ feature)-matrix: $\boldsymbol{\Phi}$

They share the same feature set, while instances are different.

In this chapter, we consider this type of situations, in which we have multiple datasets, particularly matrices, which share one or two dimensions each other. Again currently we have a variety of data in marketing, which are likely to share the same dimension(s), like instances (for example, users) or features (like in

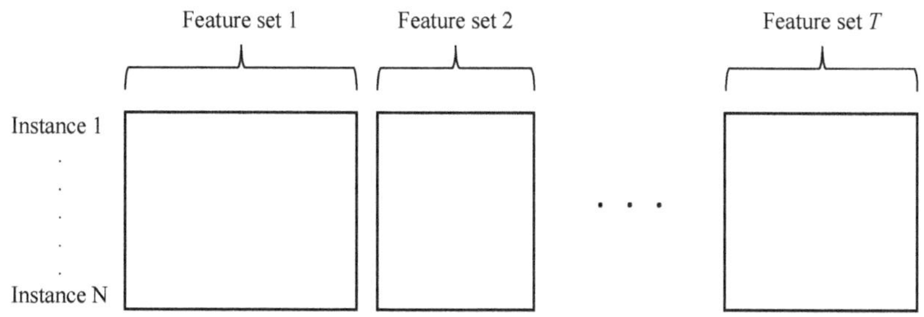

Figure 8.2: Multiview learning: schematic diagram of any arbitrary number, say T, of views, which share the same set of instances, while each of these sets of features has an unique set of features.

Section 6.4.3). We then show machine learning methods for this type of data as data integrative approaches.

8.1 Matrices (without Similarity Matrices) Sharing Instances

Just before this chapter, in Section 7.3.1, we showed an example that we have two matrices sharing the same set of instances, i.e. a (users $\times$ communication channels)-matrix and a (users $\times$ items (products))-matrix.

Generally we can explain this situation in the following way: two matrices sharing one dimension, corresponding to instances, while the other dimension of each matrix is different from each other. We write these two matrices $\boldsymbol{X}$ and $\boldsymbol{Y}$, where the rows of these two matrices are shared as instances, while columns of these two matrices are features, which are different even in their sizes. In this case, we call one set of features a *view*, and learning with more than one set of features is called *multiview learning*. Fig. 8.1 shows a schematic picture of multiview learning with two views, i.e. two matrices sharing the same set of instances but having two different sets of features. Below we introduce rather simple approaches which analyze these two matrices, under four different paradigms of machine learning:

1. **Supervised learning**

2. **Unsupervised learning: Clustering**

3. **Unsupervised learning: Matrix factorization**

4. **Learning interactions between features**

We then, under each paradigm of machine learning, further consider a general setting, which has not only two matrices, but also any arbitrary number, say R, matrices, which all share the same set of instances, while the feature set of each matrix is specific. In this case, let $\boldsymbol{X}^{(1)}, \dots, \boldsymbol{X}^{(R)}$ be R matrices (views), sharing the same instance set.

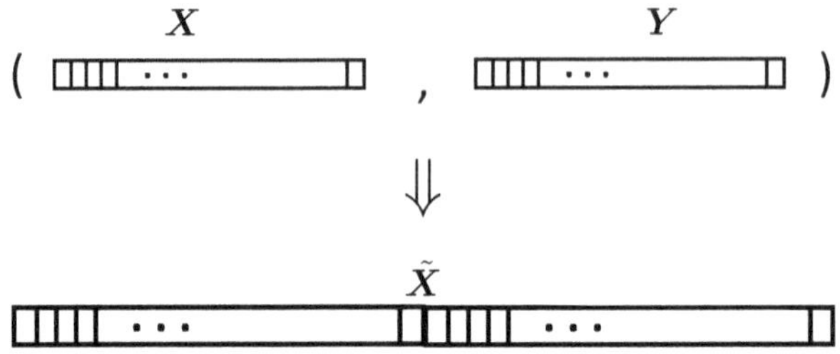

Figure 8.3: Two vectors in $\boldsymbol{X}$ and $\boldsymbol{Y}$ can be simply concatenated into a vector in $\tilde{\boldsymbol{X}}$.

Fig. 8.2 shows a schematic picture of R views sharing the same instances and having a unique set of features.

8.1.1 Supervised Learning

Two Matrices Sharing Instances

Let $\boldsymbol{q}$ and $\boldsymbol{r}$ are the label vectors of $\boldsymbol{X}$ and $\boldsymbol{Y}$, respectively. We can consider two different problem settings, depending on if $\boldsymbol{q}$ and $\boldsymbol{r}$ are the same vector or not.

1. $\boldsymbol{q} = \boldsymbol{r}$

If $\boldsymbol{q} = \boldsymbol{r}$, we can set up a simple problem setting to predict label $\boldsymbol{q}$ (i.e. $\boldsymbol{r}$) by using $\boldsymbol{X}$ or both $\boldsymbol{X}$ and $\boldsymbol{Y}$. For example, if we use both $\boldsymbol{X}$ and $\boldsymbol{Y}$ and also linear regression, parameters of linear regression are coefficients over features, which can deal with $\boldsymbol{X}$ and $\boldsymbol{Y}$ at the same time. Thus we can first simply concatenate $\boldsymbol{X}$ and $\boldsymbol{Y}$, keeping the rows the same. Fig. 8.3 shows a schematic picture of generating a vector in $\tilde{\boldsymbol{X}}$ by concatenating two vectors, one from $\boldsymbol{X}$ and the other in $\boldsymbol{Y}$, keeping the same instance. We write the concatenation in Fig. 8.3 for all instances (rows) in $\boldsymbol{X}$ and $\boldsymbol{Y}$, as follows:

$$\tilde{\boldsymbol{X}} \xLeftarrow{\text{concatenation}} \boldsymbol{X} + \boldsymbol{Y}. \tag{8.1}$$

Then we can just consider a general linear regression problem:

$$\boldsymbol{q} \sim \tilde{\boldsymbol{X}}\boldsymbol{\beta}, \tag{8.2}$$

where $\boldsymbol{\beta}$ is parameter weights over all features of $\tilde{\boldsymbol{X}}$. Thus we can formulate the problem of estimating $\boldsymbol{\beta}$, considering the constraint of $\boldsymbol{\beta}$, for example ridge regression, as follows:

$$\min_{\boldsymbol{\beta}}(\boldsymbol{q} - \tilde{\boldsymbol{X}}\boldsymbol{\beta})^{\mathsf{T}}(\boldsymbol{q} - \tilde{\boldsymbol{X}}\boldsymbol{\beta}) \text{ subject to } \lambda\boldsymbol{\beta}^{\mathsf{T}}\boldsymbol{\beta} = \text{Constant} \tag{8.3}$$

We can generate a Lagrangian from this formulation, and by setting the partial derivative of the Lagrangian with respect to β to zero, we can have the following analytical solution:

$$\hat{\boldsymbol{\beta}} = (\tilde{\boldsymbol{X}}^\top \tilde{\boldsymbol{X}} + \lambda \boldsymbol{I})^{-1} \tilde{\boldsymbol{X}}^\top \boldsymbol{q}, \tag{8.4}$$

where $\boldsymbol{I}$ is the identity matrix.

In fact this just follows a general solution of linear regression, for which $\boldsymbol{X}$ is replaced with $\tilde{\boldsymbol{X}}$. If we use $\boldsymbol{X}$ only (or $\boldsymbol{Y}$ only) instead of both $\boldsymbol{X}$ and $\boldsymbol{Y}$, then in (8.4), simply we can just replace $\tilde{\boldsymbol{X}}$ with $\boldsymbol{X}$ (or $\boldsymbol{Y}$).

2. $\boldsymbol{q} \neq \boldsymbol{r}$

We cannot use the same parameters for the two different sets of labels. Thus, instead of the shared parameter $\boldsymbol{\beta}$, we have two different parameter vectors: $\boldsymbol{\beta_X}$ and $\boldsymbol{\beta_Y}$ for two label vectors: $\boldsymbol{q}$ and $\boldsymbol{r}$, respectively. A baseline idea is that we just set up two supervised problems separately:

$$\boldsymbol{q} \sim \tilde{\boldsymbol{X}} \boldsymbol{\beta_X}, \quad \boldsymbol{r} \sim \tilde{\boldsymbol{Y}} \boldsymbol{\beta_Y}. \tag{8.5}$$

Then we estimate $\boldsymbol{\beta_X}$ and $\boldsymbol{\beta_Y}$, also separately, as follows:

$$\hat{\boldsymbol{\beta}}_{\boldsymbol{X}} = (\boldsymbol{X}^\top \boldsymbol{X} + \lambda \boldsymbol{I})^{-1} \boldsymbol{X}^\top \boldsymbol{q}, \quad \hat{\boldsymbol{\beta}}_{\boldsymbol{Y}} = (\boldsymbol{Y}^\top \boldsymbol{Y} + \lambda \boldsymbol{I})^{-1} \boldsymbol{Y}^\top \boldsymbol{r}. \tag{8.6}$$

Then even if we use both $\boldsymbol{X}$ and $\boldsymbol{Y}$, i.e. $\tilde{\boldsymbol{X}}$, what we can do is the same as the baseline method. That is, in (8.6), both $\boldsymbol{X}$ and $\boldsymbol{Y}$ are replaced with $\tilde{\boldsymbol{X}}$. Interestingly and reasonably, if we use $\tilde{\boldsymbol{X}}$, two analytical solutions in (8.6) become the same solution, except the last part, i.e. $\boldsymbol{q}$ and $\boldsymbol{r}$.

Generalization to R Matrices, Sharing Instances

If we extend the two matrices to R matrices, $\boldsymbol{X}^{(1)}, \ldots, \boldsymbol{X}^{(R)}$, which share the set of instances, we can simply use the same procedures for the above both cases. That is, if $\boldsymbol{q} = \boldsymbol{r}$, we can concatenate R matrices into $\tilde{\boldsymbol{X}}$ as follows:

$$\tilde{\boldsymbol{X}} \xLeftarrow{\text{concatenation}} \boldsymbol{X}^{(1)} + \cdots + \boldsymbol{X}^{(R)}. \tag{8.7}$$

Then we can follow the above procedure of linear regression.

If $\boldsymbol{q} \neq \boldsymbol{r}$, we can estimate each parameter vector corresponding to a given matrix independently if we deal with given T matrices independently; otherwise we can estimate a single parameter vector over given T matrices by using $\tilde{\boldsymbol{X}}$ which can be computed from (8.7).

8.1.2 Unsupervised Learning: Clustering

Two Matrices Sharing Instances

As unsupervised learning, we focus on clustering and matrix factorization. Then first we consider clustering: Each of the two sets of features can generate clusters independently, while the resultant clusters should be shared between the given two matrices, because the two matrices have the same instances. Thus we have common cluster assignment matrix $\boldsymbol{Z}$ (over both $\boldsymbol{X}$ and $\boldsymbol{Y}$), which has the same properties as those in previous sections of using constrained K-means clustering or spectral clustering.

A baseline idea is that the concatenation of $\boldsymbol{X}$ and $\boldsymbol{Y}$ to generate $\tilde{\boldsymbol{X}}$, as shown in Fig. 8.3. Then clustering, say constrained K-means clustering, can be run over $\tilde{\boldsymbol{X}}$, where the formulation can be written as follows:

$$\min_{\boldsymbol{Z}} \quad \text{trace}(\boldsymbol{Z}^\top \tilde{\boldsymbol{E}} \boldsymbol{Z}) \text{ subject to } \boldsymbol{Z}^\top \boldsymbol{Z} = \boldsymbol{D}, \tag{8.8}$$

where $\tilde{\boldsymbol{E}}$ is given as follows:

$$\tilde{\boldsymbol{E}}_{ij} = ||\tilde{\boldsymbol{x}}_i - \tilde{\boldsymbol{x}}_j||^2, \tag{8.9}$$

where $\tilde{\boldsymbol{x}}_i$ is the i-th instance of $\tilde{\boldsymbol{X}}$, and also $\boldsymbol{D}$ is a diagonal matrix.

The feature values would be, of course, different between $\boldsymbol{X}$ and $\boldsymbol{Y}$, particularly in the range and distribution of feature values. These differences were not a big problem in linear regression, which, as parameters, uses feature coefficients that can be trained from data, according to the significance of features, by which features from different matrices can be dealt with by linear regression without any problems. However the differences in features between $\boldsymbol{X}$ and $\boldsymbol{Y}$ cannot be properly considered by constrained K-means clustering, because this clustering uses similarities between instances and the difference between $\boldsymbol{X}$ and $\boldsymbol{Y}$ would not be considered by similarity computation from concatenation, i.e. $\tilde{\boldsymbol{X}}$. Thus the concatenation of two matrices would be too rude for constrained K-means clustering, and thus we need another more-acceptable approach.

A more reasonable approach would be to generate the objective function from each matrix, for example:

$$\text{trace}(\boldsymbol{Z}^\top \boldsymbol{E}^{\boldsymbol{X}} \boldsymbol{Z}), \quad \text{trace}(\boldsymbol{Z}^\top \boldsymbol{E}^{\boldsymbol{Y}} \boldsymbol{Z}), \tag{8.10}$$

where the (i, j) element of $\boldsymbol{E}^{\boldsymbol{X}}$ and $\boldsymbol{E}^{\boldsymbol{Y}}$ are

$$\boldsymbol{E}^{\boldsymbol{X}}_{ij} = ||\boldsymbol{x}_i - \boldsymbol{x}_j||^2, \quad \boldsymbol{E}^{\boldsymbol{Y}}_{ij} = ||\boldsymbol{y}_i - \boldsymbol{y}_j||^2, \tag{8.11}$$

respectively, and combine the two matrices (sets of features) at the level of the objective functions, since these two matrices should be minimized in optimization. For example, we can think about weights, $w_{\boldsymbol{X}}$ and $w_{\boldsymbol{Y}}$, over the two matrices, as follows:

$$w_{\boldsymbol{X}} \text{trace}(\boldsymbol{Z}^\top \boldsymbol{E}^{\boldsymbol{X}} \boldsymbol{Z}), \quad w_{\boldsymbol{Y}} \text{trace}(\boldsymbol{Z}^\top \boldsymbol{E}^{\boldsymbol{Y}} \boldsymbol{Z}), \tag{8.12}$$

and then combine these two weighted terms.

However if we think weights over the two matrices, each weight should show the importance of the corresponding matrix, and weights should be larger for more important matrices. This means that as we already explained in Section 6.2.1, in weight learning, the problem should be a maximization problem, while

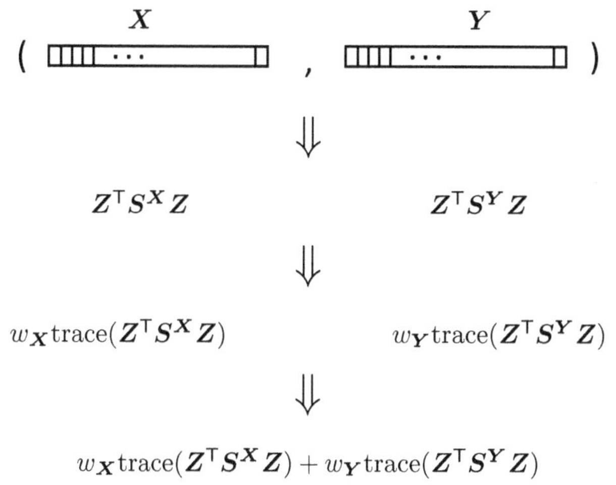

Figure 8.4: A typical example of integrating two matrices (even two different types of datasets) by using their objective functions.

the original constrained K-means clustering is a minimization problem, because of thinking the distance (variance) between instances as in (8.11).

Thus we need replace the distance in (8.11) with a similarity, which changes (8.12) as follows:

$$w_{\boldsymbol{X}}\text{trace}(\boldsymbol{Z}^\top \boldsymbol{S}^{\boldsymbol{X}} \boldsymbol{Z}), \quad w_{\boldsymbol{Y}}\text{trace}(\boldsymbol{Z}^\top \boldsymbol{S}^{\boldsymbol{Y}} \boldsymbol{Z}), \tag{8.13}$$

where we can use, as the simplest example, cosine similarity for $\boldsymbol{S}^{\boldsymbol{X}}$ and $\boldsymbol{S}^{\boldsymbol{Y}}$, as follows:

$$\boldsymbol{S}^{\boldsymbol{X}} = \boldsymbol{X}\boldsymbol{X}^\top, \quad \boldsymbol{S}^{\boldsymbol{Y}} = \boldsymbol{Y}\boldsymbol{Y}^\top. \tag{8.14}$$

Then by using cosine similarities, from **Proposition 6.2**, the following two inequalities hold:

$$\text{trace}(\boldsymbol{Z}^\top \boldsymbol{S}^{\boldsymbol{X}} \boldsymbol{Z}) \geq 0, \quad \text{trace}(\boldsymbol{Z}^\top \boldsymbol{S}^{\boldsymbol{Y}} \boldsymbol{Z}) \geq 0. \tag{8.15}$$

Then the following terms would be the objective function, which should be maximized:

$$w_{\boldsymbol{X}}\text{trace}(\boldsymbol{Z}^\top \boldsymbol{S}^{\boldsymbol{X}} \boldsymbol{Z}) + w_{\boldsymbol{Y}}\text{trace}(\boldsymbol{Z}^\top \boldsymbol{S}^{\boldsymbol{Y}} \boldsymbol{Z}). \tag{8.16}$$

Fig. 8.4 shows a schematic picture of the above procedure of integrating two matrices at the level of the objective function for the maximization problem, instead of simpler data concatenation.

Generalization to R Matrices, Sharing Instances

Furthermore this formulation can be generalized for any arbitrary number, say R, of matrices (views). For the input of these R views, we can write the objective

Procedure 8.1: INTEGRATIVE CLUSTERING FROM MULTIPLE MATRICES SHARING INSTANCES.

Input: R matrices sharing instances: $\boldsymbol{X}^{(r)} (r = 1, \ldots, R)$
Output: Cluster assignment matrix $\boldsymbol{Z}$; weights over feature groups $\boldsymbol{w}$

Preparation: compute $\boldsymbol{S}^{(r)} (r = 1, \ldots, R)$ from given $\boldsymbol{X}^{(r)} (r = 1, \ldots, R)$ through cosine similarities.;
repeat
  /* Estimate $\boldsymbol{Z}$, fixing $\boldsymbol{w}$. */
  Compute $\boldsymbol{S}$ by (8.22).;
  Estimate $\hat{\boldsymbol{Z}}$ by solving the eigenvalue problem given by (8.21).;
  Run `ClusteringPostProcessing` ($\hat{\boldsymbol{Z}}$).;
  /* Estimate $\boldsymbol{w}$, fixing $\boldsymbol{Z}$. */
  Estimate $\hat{\boldsymbol{w}}$ by running (8.23).;
until *convergence*;
Output finally estimated $\hat{\boldsymbol{Z}}$ and $\hat{\boldsymbol{w}}$.;

function of multiview clustering which integrates the R views, which for example are shown in Fig. 8.2 schematically, as follows:

$$\sum_r w_r \mathrm{trace}(\boldsymbol{Z}^\top \boldsymbol{S}^{(r)} \boldsymbol{Z}), \tag{8.17}$$

where w_r is the weight over the r-th matrix, and $\boldsymbol{S}^{(r)}$ is obtained as, for example, cosine similarities over instances in the r-th matrix:

$$\boldsymbol{S}^{(r)} = \boldsymbol{X}^{(r)} (\boldsymbol{X}^{(r)})^\top \quad (r = 1, \cdots, R) \tag{8.18}$$

Note that this is a general formulation of weight learning, which was introduced in Section 6.2.1.

We then now two types of parameters, cluster assignment matrix $\boldsymbol{Z}$ and weights over two matrices, $\boldsymbol{w} = (w_1, \ldots, w_R)^\top$. The $\boldsymbol{w}$ might be given manually, by using some prior knowledge. However we can estimate both $\boldsymbol{Z}$ and $\boldsymbol{w}$ from given data, i.e. R views.

First we consider the following constraints on $\boldsymbol{Z}$ and $\boldsymbol{w}$:

$$\boldsymbol{Z}^\top \boldsymbol{Z} = \boldsymbol{D} \quad \text{and} \quad ||\boldsymbol{w}||^2 = 1. \tag{8.19}$$

Then we can formulate our problem into the following maximization problem, using L^2 norm for constraints:

$$\max_{\boldsymbol{Z},\boldsymbol{w}} \sum_r w_r \mathrm{trace}(\boldsymbol{Z}^\top \boldsymbol{S}^{(r)} \boldsymbol{Z}) \;\; \text{subject to} \;\; \boldsymbol{Z}^\top \boldsymbol{Z} = \boldsymbol{D} \;\; \text{and} \;\; ||\boldsymbol{w}||^2 = 1. \tag{8.20}$$

In fact again this is the same formulation as that in Section 6.2.1 for clustering over groups of features. In Section 6.2.1, the assumption was features are disjointly separated into several groups each other. Instead of groups, under the current assumption, we have clearly different sets of features, resulting in the same formulation, i.e. weight learning.

This is a biconvex problem, for which a solution is to alternately 1) estimate $\boldsymbol{Z}$, assuming that $\boldsymbol{w}$ is fixed, and 2) estimate $\boldsymbol{w}$, assuming that $\boldsymbol{Z}$ is fixed.

The way of estimating each parameter is obtained by setting the partial derivative of the Lagrangian with respect to this parameter to zero. Below we show the parameter estimation manner briefly (see Section 6.2.1 for the detail of the derivation).

First $\boldsymbol{Z}$ can be estimated by solving the following eigenvalue problem:

$$\boldsymbol{S}\boldsymbol{Z} = \lambda\boldsymbol{Z}, \tag{8.21}$$

where $\boldsymbol{S}$ can be computed as follows:

$$\boldsymbol{S} \quad = \quad \sum_r w_r \boldsymbol{S}^{(r)}, \tag{8.22}$$

and the subsequent postprocessing procedure of K-means clustering over the $\boldsymbol{Z}$ obtained by solving the eigenvalue problem.

On the other hand, according to (6.38), weight $\boldsymbol{w}$ can be obtained as follows:

$$\hat{w}_r = \frac{\mathrm{trace}(\boldsymbol{Z}^\top \boldsymbol{S}^{(r)} \boldsymbol{Z})}{\sqrt{\sum_r (\mathrm{trace}(\boldsymbol{Z}^\top \boldsymbol{S}^{(r)} \boldsymbol{Z}))^2}}. \tag{8.23}$$

Overall **Procedure 8.1** shows a pseudocode of the above procedure on integrative clustering for multiple matrices sharing instances.

8.1.3 Unsupervised Learning: Matrix Factorization (Collaborative Matrix Factorization)

.

Two Matrices Sharing Instances

A baseline approach of matrix factorization for $\boldsymbol{X}$ and $\boldsymbol{Y}$ is that each of $\boldsymbol{X}$ and $\boldsymbol{Y}$ is factorized into two low-rank matrices, independently, meaning that four different low-rank matrices are generated. However, this baseline idea does not take advantage of the property of given data, in which instances are shared by two given matrices, $\boldsymbol{X}$ and $\boldsymbol{Y}$. Thus a more reasonable approach of matrix factorization would be that the low-rank matrices from $\boldsymbol{X}$ and those from $\boldsymbol{Y}$ should be (partially) shared with each other. Nicely the original matrices share the instance side, by which the low-rank matrices also should share the instance side. Fig. 8.5 shows the matrix factorization of $\boldsymbol{X}$ and $\boldsymbol{Y}$, generating three low-rank matrices, in which the instance side low-rank matrix is shared by the matrix factorization of both $\boldsymbol{X}$ and $\boldsymbol{Y}$. In general, this type of matrix factorization which

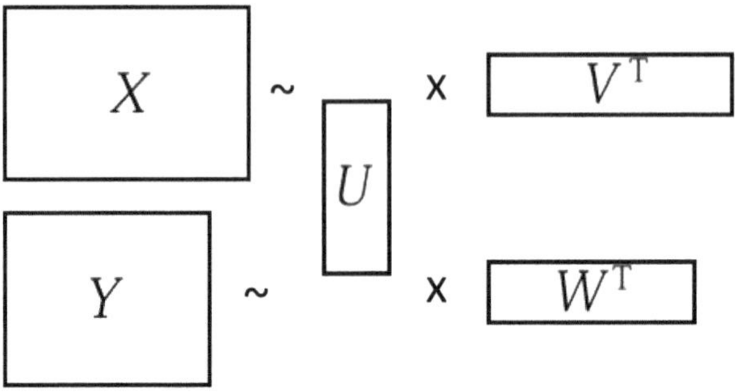

Figure 8.5: Collaborative matrix factorization, which factorizes given two matrices into three low-rank matrices.

receives more than one matrix as input and generates one or more shared low-rank matrices is called *collaborative matrix factorization*[1] [115].

As shown in Fig. 8.5, let $\boldsymbol{U}$ be the low-rank matrix shared by the matrix factorization of the both sides, and $\boldsymbol{V}$ and $\boldsymbol{W}$ be the low-rank matrix from $\boldsymbol{X}$ and $\boldsymbol{Y}$, respectively. We can then write the matrix factorization of $\boldsymbol{X}$ and $\boldsymbol{Y}$, as follows:

$$\boldsymbol{X} \sim \boldsymbol{U}\boldsymbol{V}^\top, \quad \boldsymbol{Y} \sim \boldsymbol{U}\boldsymbol{W}^\top. \tag{8.24}$$

We then consider estimating these three low-rank matrices so that we can reproduce $\boldsymbol{X}$ and $\boldsymbol{Y}$ from them. This approximation would be the minimization problem as follows:

$$\begin{aligned} \min_{\boldsymbol{U},\boldsymbol{V},\boldsymbol{W}} \quad & ||\boldsymbol{X} - \boldsymbol{U}\boldsymbol{V}^\top||^2 + \lambda_Y ||\boldsymbol{Y} - \boldsymbol{U}\boldsymbol{W}^\top||^2 \\ \text{subject to} \quad & ||\boldsymbol{U}||^2 = \text{Const.}, ||\boldsymbol{V}||^2 = \text{Const.}, ||\boldsymbol{W}||^2 = \text{Const.} \end{aligned} \tag{8.25}$$

This is a triconvex problem of three parameters $\boldsymbol{U}$, $\boldsymbol{V}$ and $\boldsymbol{W}$, where the solution is to repeat the following three steps alternately until convergence, where the converged solution is a local optimum of the optimization problem given in (8.33).

a) estimate $\boldsymbol{U}$, fixing $\boldsymbol{V}$ and $\boldsymbol{W}$.

b) estimate $\boldsymbol{V}$, fixing $\boldsymbol{U}$ and $\boldsymbol{W}$.

c) estimate $\boldsymbol{W}$, fixing $\boldsymbol{U}$ and $\boldsymbol{V}$.

In order to derive the update rule for each step, the Lagrangian for the formulation

[1] Collaborative filtering indicates prediction through the collaboration of similar users (customers), while collaborative matrix factorization indicates the collaboration amount given multiple matrices.

Procedure 8.2: COLLABORATIVE MATRIX FACTORIZATION FOR TWO MATRICES SHARING INSTANCES.

Input: Two matrices sharing instances: $\boldsymbol{X}$ and $\boldsymbol{Y}$; low-rank: K
Output: Low-rank matrices: $\boldsymbol{U}$, $\boldsymbol{V}$, $\boldsymbol{W}$

Initialization: Initialize $\boldsymbol{U}$, $\boldsymbol{V}$ and $\boldsymbol{W}$.;
repeat
  Update $\boldsymbol{U}$ by (8.30).;
  Update $\boldsymbol{V}$ by (8.31).;
  Update $\boldsymbol{W}$ by (8.32).;
until *convergence*;
Output finally estimated $\hat{\boldsymbol{U}}$, $\hat{\boldsymbol{V}}$ and $\hat{\boldsymbol{W}}$.;

given in (8.33) is given as follows:

$$L(\boldsymbol{U},\boldsymbol{V},\boldsymbol{W}) = ||\boldsymbol{X}-\boldsymbol{U}\boldsymbol{V}^\top||^2 + \lambda_Y||\boldsymbol{Y}-\boldsymbol{U}\boldsymbol{W}^\top||^2 + \lambda(||\boldsymbol{U}||^2 + ||\boldsymbol{V}||^2 + ||\boldsymbol{W}||^2). \tag{8.26}$$

By setting the partial derivative of the above Lagrangian with respect to each of the three parameters to zero, we can obtain the following additive update rule for estimating each of the three parameters[2]:

$$\frac{\partial L(\boldsymbol{U},\boldsymbol{V},\boldsymbol{W})}{\partial \boldsymbol{U}} = \boldsymbol{0} \Rightarrow \boldsymbol{U} \leftarrow (\boldsymbol{X}\boldsymbol{V} + \lambda_Y \boldsymbol{Y}\boldsymbol{W})(\boldsymbol{V}^\top\boldsymbol{V} + \lambda_Y \boldsymbol{W}^\top\boldsymbol{W} + \lambda\boldsymbol{I})^{-1}, \tag{8.27}$$

$$\frac{\partial L(\boldsymbol{U},\boldsymbol{V},\boldsymbol{W})}{\partial \boldsymbol{V}} = \boldsymbol{0} \Rightarrow \boldsymbol{V}^\top \leftarrow (\boldsymbol{U}^\top\boldsymbol{U} + \lambda\boldsymbol{I})^{-1}\boldsymbol{U}^\top\boldsymbol{X}, \tag{8.28}$$

$$\frac{\partial L(\boldsymbol{U},\boldsymbol{V},\boldsymbol{W})}{\partial \boldsymbol{W}} = \boldsymbol{0} \Rightarrow \boldsymbol{W}^\top \leftarrow (\boldsymbol{U}^\top\boldsymbol{U} + \lambda_W\boldsymbol{I})^{-1}\boldsymbol{U}^\top\boldsymbol{Y}, \tag{8.29}$$

where $\lambda_W = \frac{\lambda}{\lambda_Y}$.

We can first easily see that the additive update rules of $\boldsymbol{V}$ and $\boldsymbol{W}$, given by (8.28) and (8.29), respectively, are the same as that of $\boldsymbol{V}$ of bimatrix factorization, given in (7.23). On the other hand, the update rule of $\boldsymbol{U}$ is different from that of the bimatrix factorization, given in (7.22). Also we can see that the update of $\boldsymbol{U}$ is affected by both $\boldsymbol{X}$ and $\boldsymbol{Y}$. On the other hand, the update of $\boldsymbol{V}$ ($\boldsymbol{W}$) depends on $\boldsymbol{U}$ and $\boldsymbol{X}$ ($\boldsymbol{Y}$) only, having nothing to do with $\boldsymbol{Y}$ ($\boldsymbol{X}$). However $\boldsymbol{U}$ is affected by the both sides, meaning that the effects of $\boldsymbol{Y}$ ($\boldsymbol{X}$) on $\boldsymbol{V}$ ($\boldsymbol{W}$) are propagated through $\boldsymbol{U}$.

Similarly the multiplicative update rules for estimating each of the three pa-

[2]We omit the derivation, since these updating rules can be understood through the analogy of only one input matrix.

rameters can be obtained as follows[3]:

$$\boldsymbol{U}_{ik} \leftarrow \boldsymbol{U}_{ik}\frac{(\boldsymbol{X}\boldsymbol{V}+\lambda_Y\boldsymbol{Y}\boldsymbol{W})_{ik}}{(\boldsymbol{U}(\boldsymbol{V}^\mathsf{T}\boldsymbol{V}+\lambda_Y\boldsymbol{W}^\mathsf{T}\boldsymbol{W}+\lambda\boldsymbol{I}))_{ik}}, \tag{8.30}$$

$$\boldsymbol{V}_{jk} \leftarrow \boldsymbol{V}_{jk}\frac{(\boldsymbol{X}^\mathsf{T}\boldsymbol{U})_{jk}}{(\boldsymbol{V}(\boldsymbol{U}^\mathsf{T}\boldsymbol{U}+\lambda\boldsymbol{I}))_{jk}}, \tag{8.31}$$

$$\boldsymbol{W}_{jk} \leftarrow \boldsymbol{W}_{jk}\frac{(\boldsymbol{Y}^\mathsf{T}\boldsymbol{U})_{jk}}{(\boldsymbol{W}(\boldsymbol{U}^\mathsf{T}\boldsymbol{U}+\lambda_W\boldsymbol{I}))_{jk}}, \tag{8.32}$$

where again $\lambda_W = \frac{\lambda}{\lambda_Y}$.

Again the update of $\boldsymbol{V}$ ($\boldsymbol{W}$) is kept the same as that in (7.27). Also similar to the additive update rule (8.27), the effect of $\boldsymbol{Y}$ ($\boldsymbol{X}$) on parameter $\boldsymbol{V}$ is conveyed through $\boldsymbol{U}$, which is affected by all other matrices, i.e. $\boldsymbol{X}$, $\boldsymbol{Y}$, $\boldsymbol{U}$ and $\boldsymbol{V}$. Thus we can say that collaborative matrix factorization has an interesting property by sharing the low-rank matrix from given multiple matrices. Also among the three low-rank matrices, we can regard $\boldsymbol{U}$ as the matrix with the shared property of given input $\boldsymbol{X}$ and $\boldsymbol{Y}$, while $\boldsymbol{V}$ ($\boldsymbol{W}$) is regarded as the matrix keeping the property specific to $\boldsymbol{X}$ ($\boldsymbol{Y}$).

Finally **Procedure 8.2** shows a pseudocode of the procedure for estimating the three low-rank matrices of the collaborative matrix factorization of $\boldsymbol{X}$ and $\boldsymbol{Y}$.

Generalization to R Matrices, Sharing Instances

Also again, we can generalize this collaborative matrix factorization for given two matrices (views) to that for any arbitrary number, say R, of matrices, i.e. multi-views, which share the same set of instances but have feature sets, in which each set is specific to one view. Fig. 8.6 shows a schematic diagram of this generalized collaborative matrix factorization with R views.

We here show the method to solve the problem of estimating $R+1$ low-rank matrices from the input R matrices. Let $\boldsymbol{X}^{(1)}, \ldots, \boldsymbol{X}^{(R)}$ be R input matrices, which share the same set of instances and differ in the set of features. Let $\boldsymbol{U}$ be the shared low-rank matrix among the R input matrices, as shown in Fig. 8.6. Let $\boldsymbol{V}^{(1)}, \ldots, \boldsymbol{V}^{(R)}$ be R low-rank matrices, where $\boldsymbol{V}^{(r)}$ corresponds to $\boldsymbol{X}^{(r)}$.

Obviously this is an extension of bi- or triconvex problem, i.e. *multiconvex problem*. Then we can formulate the problem of estimating the low-rank matrices which are factorized from the input and also can reproduce the original input, as the following minimization problem:

$$\begin{aligned} \min_{\boldsymbol{U},\boldsymbol{V}^{(1)},\ldots,\boldsymbol{V}^{(R)}} \quad & ||\boldsymbol{X}^{(1)} - \boldsymbol{U}(\boldsymbol{V}^{(1)})^\mathsf{T}||^2 + \cdots + \lambda_R||\boldsymbol{X}^{(R)} - \boldsymbol{U}(\boldsymbol{V}^{(R)})^\mathsf{T}||^2 \\ \text{subject to} \quad & ||\boldsymbol{U}||^2 = \text{Const.}, ||\boldsymbol{V}^{(1)}||^2 = \text{Const.}, \ldots, ||\boldsymbol{V}^{(R)}||^2 = \text{Const.} \end{aligned} \tag{8.33}$$

In this optimization problem, we have $R+1$ matrix parameters, and one matrix parameter can be estimated by solving a convex problem, by fixing other R matrix

[3]We omit the derivation of the multiplicative rules as well.

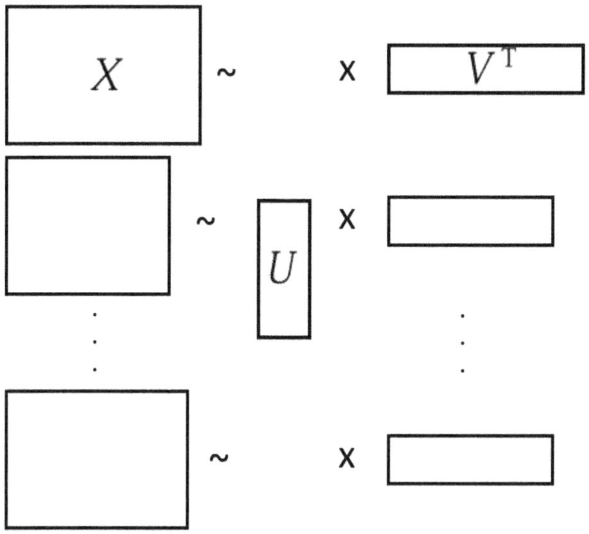

Figure 8.6: Collaborative matrix factorization, which factorizes given R matrices into $R+1$ low-rank matrices, where one low-rank matrix is shared and each of the other R low-rank matrices is specific to the corresponding given matrix.

parameters. This means that we repeat the following $R+1$ steps alternately:

1) estimate $\boldsymbol{U}$, fixing $\boldsymbol{V}^{(1)},\ldots,\boldsymbol{V}^{(R)}$.

2) estimate $\boldsymbol{V}^{(1)}$, fixing $\boldsymbol{U}$ and $\boldsymbol{V}^{(2)},\ldots,\boldsymbol{V}^{(R)}$.

.

.

.

$R+1$) estimate $\boldsymbol{V}^{(R)}$, fixing $\boldsymbol{U}$ and $\boldsymbol{V}^{(1)},\ldots,\boldsymbol{V}^{(R-1)}$.

We then have the following Lagrangian function from (8.33), using L^2 norm for the constraints.

$$\begin{aligned} L(\boldsymbol{U},\boldsymbol{V}^{(1)},\ldots,\boldsymbol{V}^{(R)}) &= ||\boldsymbol{X}-\boldsymbol{U}(\boldsymbol{V}^{(1)})^{\mathsf{T}}||^2+\cdots+\lambda_R||\boldsymbol{X}-\boldsymbol{U}(\boldsymbol{V}^{(R)})^{\mathsf{T}}||^2 \\ &+ \lambda(||\boldsymbol{U}||^2+||\boldsymbol{V}^{(1)}||^2+\cdots+||\boldsymbol{V}^{(R)}||^2). \end{aligned} \quad (8.34)$$

By setting the partial derivative of this Lagrangian with respect to each parameter to zero, we can derive the following additive update rule for each of the

---

Procedure 8.3: COLLABORATIVE MATRIX FACTORIZATION FOR T MATRICES SHARING INSTANCES.

Input: R matrices sharing instances: $\boldsymbol{X}^{(1)}, \ldots, \boldsymbol{X}^{(R)}$; low-rank: K
Output: Low-rank matrices: $\boldsymbol{U}$, $\boldsymbol{V}^{(1)}, \ldots, \boldsymbol{V}^{(R)}$

```
Initialization: Initialize U, V^(1), ..., V^(R).;
repeat
    Update U by (8.37).;
    for r ← 1 to R do
        Update V^(r) by (8.38).;
until convergence;
Output finally estimated Û and V̂^(r) (r = 1, ..., R).;
```

$R+1$ parameters[4]:

$$\frac{\partial L(\boldsymbol{U}, \boldsymbol{V}^{(1)}, \ldots, \boldsymbol{V}^{(R)})}{\partial \boldsymbol{U}} = \boldsymbol{0} \quad \Rightarrow \quad \boldsymbol{U} \leftarrow (\boldsymbol{X}^{(1)}\boldsymbol{V}^{(1)} + \cdots + \lambda_R \boldsymbol{X}^{(R)}\boldsymbol{V}^{(R)}) \\ ((\boldsymbol{V}^{(1)})^\top \boldsymbol{V}^{(1)} + \cdots + \lambda_R (\boldsymbol{V}^{(R)})^\top \boldsymbol{V}^{(R)} + \lambda \boldsymbol{I})^{-1}, \tag{8.35}$$

$$\frac{\partial L(\boldsymbol{U}, \boldsymbol{V}^{(1)}, \ldots, \boldsymbol{V}^{(R)})}{\partial \boldsymbol{V}^{(r)}} = \boldsymbol{0} \quad \Rightarrow \quad (\boldsymbol{V}^{(r)})^\top \leftarrow (\boldsymbol{U}^\top \boldsymbol{U} + \gamma_r \boldsymbol{I})^{-1} \boldsymbol{U}^\top \boldsymbol{X}^{(r)} \\ (r = 1, \ldots, R), \tag{8.36}$$

where $\gamma_r = \frac{\lambda}{\lambda_r}$.

The update rules of collaboration matrix factorization with the input of R matrices have the same properties as those with the input of two matrices. As we can easily see from (8.36), the update rules of the input matrix specific parameters (i.e. $\boldsymbol{V}^{(1)}, \ldots, \boldsymbol{V}^{(R)}$) all take the same form except the last part (i.e. $\boldsymbol{X}^{(r)}$) corresponding to the input. On the other hand, in (8.35), $\boldsymbol{U}$ is updated by using all other matrices, i.e. R input matrices and R low-rank matrices.

Similarly multiplicative update rules for $R+1$ low rank matrices can be derived as follows[5]:

$$\boldsymbol{U}_{ik} \leftarrow \boldsymbol{U}_{ik} \frac{(\boldsymbol{X}^{(1)}\boldsymbol{V}^{(1)} + \cdots + \lambda_T \boldsymbol{X}^{(R)}\boldsymbol{V}^{(R)})_{ik}}{(\boldsymbol{U}((\boldsymbol{V}^{(1)})^\top \boldsymbol{V}^{(1)} + \cdots + \lambda_T (\boldsymbol{V}^{(R)})^\top \boldsymbol{V}^{(R)} + \lambda \boldsymbol{I}))_{ik}}, \tag{8.37}$$

$$\boldsymbol{V}^{(r)}_{jk} \leftarrow \boldsymbol{V}^{(r)}_{jk} \frac{((\boldsymbol{X}^{(r)})^\top \boldsymbol{U})_{jk}}{(\boldsymbol{V}^{(r)}(\boldsymbol{U}^\top \boldsymbol{U} + \gamma_r \boldsymbol{I}))_{jk}} \quad (r = 1, \ldots, R). \tag{8.38}$$

Again the multiplicative rules of collaborative matrix factorization for the input of R matrices keep the same properties as those of two matrices. That is,

---

[4]We omit the detailed derivation, while readers can easily see these update rules by the analogy from (8.27), (8.28) and (8.29).

[5]Again derivation is omitted, while readers can easily retrieve the updating rules, just from the analogy of the collaborative matrix factorization with the input of two matrices.

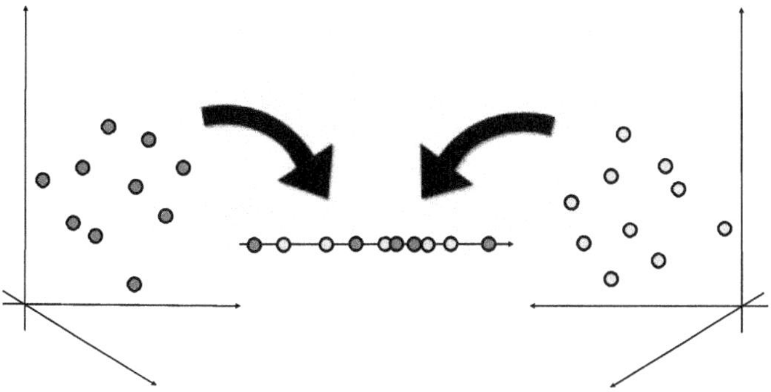

Figure 8.7: Concept of canonical correlation analysis: Two sets of instances are projected on the same 1D space. Note that we can assume different sizes for the two input matrices.

$\boldsymbol{V}^{(r)}$ can be updated by using only $\boldsymbol{U}$ and $\boldsymbol{X}^{(r)}$, i.e. the corresponding input matrix, while $\boldsymbol{U}$ is updated by all other information, i.e. all input $\boldsymbol{X}^{(1)}, \ldots, \boldsymbol{X}^{(R)}$ and all other low-rank matrices $\boldsymbol{V}^{(1)}, \ldots, \boldsymbol{V}^{(R)}$.

Procedure 8.3 shows a pseudocode of the procedure of estimating $R+1$ low-rank matrices from the input of R matrices, which share the set of instances (while the feature sets are different from each other).

8.1.4 Examining Interactions between Features

Two Matrices Sharing Instances

We have shown three machine learning paradigms to deal with given two or more matrices, generally R matrices. For example, supervised learning can classify instances by using given R matrices, and also clustering can use an arbitrary number of matrices to do grouping instances. Matrix factorization factorizes given R input matrices into $R+1$ low-rank matrices, one with the shared information by the given matrices and each of the other R low rank matrices acquiring factors specific to each of the R input matrices. The update rule of estimating low-rank matrices of matrix factorization, showed the information on features of each input matrix is transferred to a low-rank matrix (say $\boldsymbol{V}^{(r)}$), specific to each (say $\boldsymbol{X}^{(r)}$) of the other input matrices through the low-rank matrix, $\boldsymbol{U}$, shared by all input matrices. Then now our interest is shifting to examine the relationships between two sets of features given as two different views.

We show an approach called *Canonical Correlation Analysis* (CCA) for examining the relationships between features from two matrices, following [94] and Section 9.1.2 of [70]. In fact since two matrices $\boldsymbol{X}$ and $\boldsymbol{Y}$ share the same instance set, the idea of comparing two different features is to project these features into a common space, so that we can check similarity/ correlation between features easily.

This idea is rather similar to PCA (Principal Component Analysis), which also projects instances into one dimensional (1D) space, so that we can discriminate given instances as much as possible. This means that the distribution of the projected instances should be as broad as possible, resulting in that the variance of the projected instances is maximized. Following Section 5.3.2, we first explain PCA briefly: Let $\boldsymbol{w}_X$ be coefficients (weights) over the features of given input $\boldsymbol{X}$, and we project $\boldsymbol{X}$ into a 1D space by $\boldsymbol{X}\boldsymbol{w}_X$, and the variance of the mapped instances (i.e. $(\boldsymbol{X}\boldsymbol{w}_X)^\top \boldsymbol{X}\boldsymbol{w}_X = \boldsymbol{w}_X^\top \boldsymbol{X}^\top \boldsymbol{X}\boldsymbol{w}_X$) is maximized under the L^2 norm constraint of the coefficients, as follows:

$$\max_{\boldsymbol{w}} \boldsymbol{w}_X^\top \boldsymbol{X}^\top \boldsymbol{X}\boldsymbol{w}_X \quad \text{subject to} \quad \boldsymbol{w}_X^\top \boldsymbol{w}_X = \text{Constant}. \tag{8.39}$$

The solution is obtained by the following eigenvalue problem:

$$\boldsymbol{X}^\top \boldsymbol{X}\boldsymbol{w}_X = \lambda \boldsymbol{w}_X. \tag{8.40}$$

PCA has only one matrix as input, while as input CCA has two matrices $\boldsymbol{X}$ and $\boldsymbol{Y}$, which are projected into the same 1D space by $\boldsymbol{X}\boldsymbol{w}_X$ and $\boldsymbol{Y}\boldsymbol{w}_Y$, respectively. We then maximize the covariance of these two projected sets of instances which is given as follows:

$$(\boldsymbol{X}\boldsymbol{w}_X)^\top \boldsymbol{Y}\boldsymbol{w}_Y = \boldsymbol{w}_X^\top \boldsymbol{X}^\top \boldsymbol{Y}\boldsymbol{w}_Y. \tag{8.41}$$

As well as PCA, CCA has the constraints over the coefficients, while the constraint terms are not only on the coefficient parameters but also given inputs, resulting in the objective function itself of PCA as a L^2 norm constraint, as follows:

$$(\boldsymbol{X}\boldsymbol{w}_X)^\top \boldsymbol{X}\boldsymbol{w}_X = \boldsymbol{w}_X^\top \boldsymbol{X}^\top \boldsymbol{X}\boldsymbol{w}_X = \text{Constant}. \tag{8.42}$$

$$(\boldsymbol{Y}\boldsymbol{w}_Y)^\top \boldsymbol{Y}\boldsymbol{w}_Y = \boldsymbol{w}_Y^\top \boldsymbol{Y}^\top \boldsymbol{Y}\boldsymbol{w}_Y = \text{Constant}. \tag{8.43}$$

Overall we can formulate the problem of CCA as follows:

$$\max_{\boldsymbol{w}_X, \boldsymbol{w}_Y} \quad \boldsymbol{w}_X^\top \boldsymbol{C}_{XY} \boldsymbol{w}_Y \tag{8.44}$$

$$\text{subject to} \quad \boldsymbol{w}_X^\top \boldsymbol{C}_{XX} \boldsymbol{w}_X = \text{Const. and } \boldsymbol{w}_Y^\top \boldsymbol{C}_{YY} \boldsymbol{w}_Y = \text{Const.}, \tag{8.45}$$

where $\boldsymbol{C}_{XX} = \boldsymbol{X}^\top \boldsymbol{X}$, $\boldsymbol{C}_{XY} = \boldsymbol{Y}^\top \boldsymbol{X}$ and $\boldsymbol{C}_{YY} = \boldsymbol{Y}^\top \boldsymbol{Y}$.

From this problem formulation, we can define the following Lagrangian:

$$L(\boldsymbol{w}_X, \boldsymbol{w}_Y) = \boldsymbol{w}_X^\top \boldsymbol{C}_{XY} \boldsymbol{w}_Y - \lambda_X (\boldsymbol{w}_X^\top \boldsymbol{C}_{XX} \boldsymbol{w}_X - 1) - \lambda_Y (\boldsymbol{w}_Y^\top \boldsymbol{C}_{YY} \boldsymbol{w}_Y - 1), \tag{8.46}$$

where λ_X and λ_Y are regularization coefficients.

Then by setting the derivative of the Lagrange function with respect to $\boldsymbol{w}_X$ ($\boldsymbol{w}_Y$) equal to zero, we can see that the optimum $\boldsymbol{w}_X$ and $\boldsymbol{w}_Y$ satisfy the following equations.

$$\frac{\partial L(\boldsymbol{w}_X, \boldsymbol{w}_Y)}{\partial \boldsymbol{w}_X} = 0 \quad \Rightarrow \quad \boldsymbol{C}_{XY} \boldsymbol{w}_Y - \lambda_X \boldsymbol{C}_{XX} \boldsymbol{w}_X = 0. \tag{8.47}$$

$$\frac{\partial L(\boldsymbol{w}_X, \boldsymbol{w}_Y)}{\partial \boldsymbol{w}_Y} = 0 \quad \Rightarrow \quad \boldsymbol{C}_{XY}^\top \boldsymbol{w}_X - \lambda_Y \boldsymbol{C}_{YY} \boldsymbol{w}_Y = 0. \tag{8.48}$$

Procedure 8.4: CANONICAL CORRELATION ANALYSIS ALGORITHM.

```
1 Function Canonical_correlation_analysis(X, Y)
    Input: X, Y
    Output: w_X, w_Y
2   Compute C_XX, C_XY and C_XY from X and Y. ;
3   Solve the generalized eigenvalue problem of CCA, i.e. (8.50), to output
    finally estimated ŵ_X and ŵ_Y.;
```

Removing the term of $\boldsymbol{C}_{XY}$, we can have the following two equations:

$$\lambda_X \boldsymbol{w}_X^\top \boldsymbol{C}_{XX} \boldsymbol{w}_X = \lambda_Y \boldsymbol{w}_Y^\top \boldsymbol{C}_{YY} \boldsymbol{w}_Y. \tag{8.49}$$

This implies $\lambda_X = \lambda_Y = \lambda$. Then from (8.47) and (8.48), we can obtain optimum $\boldsymbol{w}_X$ and $\boldsymbol{w}_Y$ by solving the following generalized eigenvalue problem:

$$\begin{pmatrix} \mathbf{0} & \boldsymbol{C}_{XY} \\ \boldsymbol{C}_{YX} & \mathbf{0} \end{pmatrix} \begin{pmatrix} \boldsymbol{w}_X \\ \boldsymbol{w}_Y \end{pmatrix} = \lambda \begin{pmatrix} \boldsymbol{C}_{XX} & \mathbf{0} \\ \mathbf{0} & \boldsymbol{C}_{YY} \end{pmatrix} \begin{pmatrix} \boldsymbol{w}_X \\ \boldsymbol{w}_Y \end{pmatrix}, \tag{8.50}$$

where note that $\boldsymbol{C}_{XY}^\top = (\boldsymbol{X}^\top \boldsymbol{Y})^\top = \boldsymbol{Y}^\top \boldsymbol{X} = \boldsymbol{C}_{YX}$.

In fact the sizes of both rows and columns of $\boldsymbol{C}_*$ are not the size of instances, which makes the computation of solving the generalized eigenvalue problem fast. The detailed description on this point is shown in [94] and Section 9.1.2 of [70]. **Procedure 8.4** shows a pseudocode of the procedure of estimating parameters of CCA.

Generalization to R Matrices, Sharing Instances

We consider the situation of having an arbitrary number, say R, of matrices, which share the same set of instances. That is, we have R views. First (8.47) and (8.48) can be transformed into the following two equations:

$$\boldsymbol{C}_{XX} \boldsymbol{w}_X + \boldsymbol{C}_{XY} \boldsymbol{w}_Y - (1 + \lambda_X) \boldsymbol{C}_{XX} \boldsymbol{w}_X = 0. \tag{8.51}$$

$$\boldsymbol{C}_{XY}^\top \boldsymbol{w}_X + \boldsymbol{C}_{YY} \boldsymbol{w}_Y - (1 + \lambda_Y) \boldsymbol{C}_{YY} \boldsymbol{w}_Y = 0. \tag{8.52}$$

This indicates that the generalized eigenvalue problem (8.50) also can be written as follows:

$$\begin{pmatrix} \boldsymbol{C}_{XX} & \boldsymbol{C}_{XY} \\ \boldsymbol{C}_{YX} & \boldsymbol{C}_{YY} \end{pmatrix} \begin{pmatrix} \boldsymbol{w}_X \\ \boldsymbol{w}_Y \end{pmatrix} = (1 + \lambda) \begin{pmatrix} \boldsymbol{C}_{XX} & \mathbf{0} \\ \mathbf{0} & \boldsymbol{C}_{YY} \end{pmatrix} \begin{pmatrix} \boldsymbol{w}_X \\ \boldsymbol{w}_Y \end{pmatrix}. \tag{8.53}$$

From this simple observation, if R different matrices (views), i.e. $\boldsymbol{X}^{(1)}, \ldots, \boldsymbol{X}^{(R)}$, are given, we can consider the following generalized eigenvalue problem, to obtain the optimum coefficients over features of each matrix, as

Procedure 8.5: GENERALIZED CANONICAL CORRELATION ANALYSIS.

```
Function Generalized_canonical_correlation_analysis(X^(1), ..., X^(R))
    Data: X^(1), ..., X^(R)
    Result: w_1, ..., w_R
    for i ← 1 to R do
        for j ← 1 to R do
            Compute C_ij = (X^(i))^T X^(j).;
    Solve the generalized eigenvalue problem of (8.54), to output finally
    estimated ŵ_1, ..., ŵ_R.;
```

follows:

$$\begin{pmatrix} C_{11} & C_{12} & \dots & C_{1R} \\ C_{21} & C_{22} & \dots & C_{2R} \\ \vdots & \vdots & \ddots & \vdots \\ C_{R1} & C_{R2} & \dots & C_{RR} \end{pmatrix} \begin{pmatrix} w_1 \\ w_2 \\ \vdots \\ w_R \end{pmatrix} = \lambda \begin{pmatrix} C_{11} & 0 & \dots & 0 \\ 0 & C_{22} & \dots & 0 \\ \vdots & \vdots & \ddots & \vdots \\ 0 & 0 & \dots & C_{RR} \end{pmatrix} \begin{pmatrix} w_1 \\ w_2 \\ \vdots \\ w_R \end{pmatrix}. \tag{8.54}$$

where $\boldsymbol{C}_{ij} = (\boldsymbol{X}^{(i)})^\top \boldsymbol{X}^{(j)}$.

Overall we call this CCA for any arbitrary number of matrices as input the *generalized canonical correlation analysis* (GCCA). **Procedure 8.5** is a pseudocode of the procedure for estimating parameters of GCCA, where the input is T matrices, $\boldsymbol{X}^{(1)}, \dots, \boldsymbol{X}^{(R)}$ and the output is R vectors of coefficients over features.

8.1.5 Notes

We briefly mention possible extensions of the machine learning methods which we have raised for multiple matrices sharing the same set of instance.

Kernel learning: First, we can apply kernel learning to each of the four methods. In fact, kernel ridge regression, kernel K-means and kernel CCA are all straightforward extensions from ridge regression, K-means clustering and CCA, respectively (see Section 3.4.2, Section 3.4.1 and Section 9.1.3 of [70] for kernel ridge regression, kernel K-means clustering and kernel CCA, respectively). Among them, kernel CCA already assumes the input of more than one matrix, while kernel ridge regression and kernel K-means are for the input of only one matrix. However, they can be extended to dealing with multiple matrices rather easily, as we have shown the extension of regular ridge regression and K-means for one matrix to those for any arbitrary number of matrices. Kernel learning has the advantage of incorporating some prior knowledge as kernels, which would be helpful for aiming higher predictive performance. On the other hand, kernel functions are assumed to be

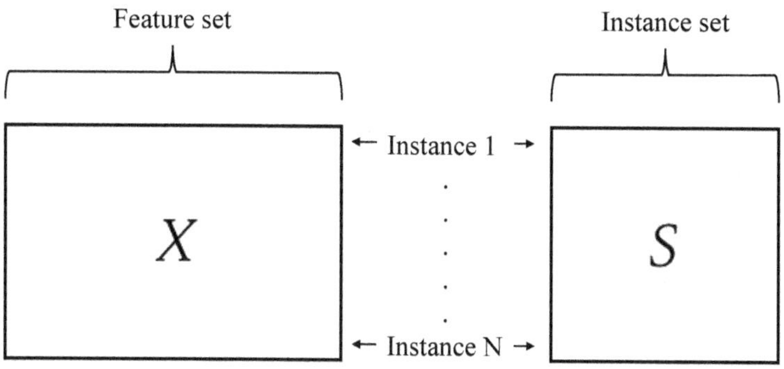

Figure 8.8: Multiview learning under a special constraint: schematic diagram of two input matrices, sharing the same set of instance, where one of the two input matrices is a similarity matrix.

no missing information, which would be a demerit, when we use real-world matrices.

8.2 Matrices (with Similarity Matrices) Sharing Instances

Now we consider the input of two matrices, where one is a regular matrix (of instances × features) while the other matrix is a similarity matrix over instances. Fig. 8.8 shows a schematic picture of this input. Note that this is a special case of the input in Section 8.1, which has two matrices sharing the set of instances. In this section, also the input is two matrices, sharing the set of instances, while the constraint of this chapter is that one of the two input matrices is a similarity matrix. Practically, a real example of this situation is that one matrix is a user-item matrix and the other matrix is on user (or item) similarity or a social network, which means a graph or an adjacency matrix.

Below we explain rather simple machine learning approaches for this input, again under three different paradigms of machine learning[6].

1. **Supervised learning**

2. **Unsupervised learning: Clustering**

3. **Unsupervised learning: Matrix factorization**

Suppose that we have two input matrices: let $\boldsymbol{X}$ and $\boldsymbol{S}$ be a regular matrix and a similarity (adjacency) matrix.

[6] In the general situation of Section 8.1, we have four paradigms, in which the last paradigm was on the interaction between two feature sets. However under the current special case, we have only one feature set, and then we are unable to consider the interaction between features sets. Thus the last paradigm in this situation, i.e. interactions between two sets of features, is omitted.

Then under each paradigm, we consider a general case with multiple views, i.e. R views: $\boldsymbol{X}^{(1)}, \ldots, \boldsymbol{X}^{(R)}$, sharing the same set of instances and additionally also multiple similarity matrices, i.e. S similarity matrices: $\boldsymbol{S}^{(1)}, \ldots, \boldsymbol{S}^{(S)}$, which also share the same set of instances for both rows and columns. In other words, this situation can be regarded as that a set of instances has R views and S similarity matrices: more simply R sets of features and S adjacency matrices.

8.2.1 Supervised Learning

Two Matrices with Similarity Matrix, Sharing Instances

Each instance has a label for supervised learning. Then let $\boldsymbol{y}$ be the given label vector and $\boldsymbol{z}$ be the parameter vector for the label vector. They should be consistent with each other:

$$\min_{\boldsymbol{z}} ||\boldsymbol{z} - \boldsymbol{y}||^2. \tag{8.55}$$

The similarity matrix is equivalent to a graph or adjacency matrix, showing the similarity between instances. Thus we can think that this situation is the same as the supervised setting in Section 5.2.3 which integrates two types of data, i.e. a vector and a node in a graph, in the frame work of linear regression. We briefly review this approach.

First due to linear regression, we have one more parameter $\boldsymbol{\beta}$, which is the weight vector (coefficients) over features in $\boldsymbol{X}$, to satisfy the following relation with $\boldsymbol{z}$:

$$\boldsymbol{z} \sim \boldsymbol{X\beta} \quad \Rightarrow \quad \min_{\boldsymbol{z},\boldsymbol{\beta}} ||\boldsymbol{z} - \boldsymbol{X\beta}||^2. \tag{8.56}$$

At the same time, $\boldsymbol{z}$ should be smooth over the given graph:

$$\min_{\boldsymbol{z}} \boldsymbol{z}^\top \boldsymbol{L} \boldsymbol{z}, \tag{8.57}$$

where $\boldsymbol{L}$ is graph Laplacian of given graph, i.e. adjacency matrix $\boldsymbol{S}$. Thus overall we can formulate this problem as a minimization problem, for which we can generate the following Lagrangian, considering the L^2 norm constraint for $\boldsymbol{\beta}$:

$$\begin{aligned} L(\boldsymbol{z}, \boldsymbol{\beta}) &= (\boldsymbol{z} - \boldsymbol{y})^\top (\boldsymbol{z} - \boldsymbol{y}) + \lambda_v (\boldsymbol{z} - \boldsymbol{X\beta})^\top (\boldsymbol{z} - \boldsymbol{X\beta}) \\ &+ \lambda_b \boldsymbol{\beta}^\top \boldsymbol{\beta} + \lambda_g \boldsymbol{z}^\top \boldsymbol{L} \boldsymbol{z}, \end{aligned} \tag{8.58}$$

where λ_v, λ_b and λ_g are hyperparameters. Then this is a biconvex problem to be solved by repeating the following two estimation alternately:

$$\frac{\partial L(\boldsymbol{z}, \boldsymbol{\beta})}{\partial \boldsymbol{z}} = 0 \quad \Rightarrow \quad \hat{\boldsymbol{z}} = ((1 + \lambda_v)\boldsymbol{I} + \lambda_g \boldsymbol{L})^{-1} (\boldsymbol{y} + \lambda_v \boldsymbol{X\beta}). \tag{8.59}$$

$$\frac{\partial L(\boldsymbol{z}, \boldsymbol{\beta})}{\partial \boldsymbol{\beta}} = 0 \quad \Rightarrow \quad \hat{\boldsymbol{\beta}} = (\boldsymbol{X}^\top \boldsymbol{X} + \lambda_\beta \boldsymbol{I})^{-1} \boldsymbol{X}^\top \boldsymbol{z}, \tag{8.60}$$

where $\lambda_\beta = \frac{\lambda_b}{\lambda_v}$.

Procedure 5.5 shows a pseudocode of the procedure we briefly mentioned in the above.

Generalization to R Matrices and S Similarity Matrices, Sharing Instances

Again regular R matrices can be written as $\boldsymbol{X}^{(1)}, \ldots, \boldsymbol{X}^{(R)}$, which share the same set of instances and R similarity matrices can be written as $\boldsymbol{S}^{(1)}, \ldots, \boldsymbol{S}^{(S)}$, which also share the same set of instances for both rows and columns.

First, for $\boldsymbol{X}^{(1)}, \ldots, \boldsymbol{X}^{(R)}$, as we described for supervised learning in Section 8.1.1, we can first concatenate all given matrices as follows:

$$\tilde{\boldsymbol{X}} \xLeftarrow{\text{concatenation}} \boldsymbol{X}^{(1)} + \cdots + \boldsymbol{X}^{(R)}, \tag{8.61}$$

Then the generated $\tilde{\boldsymbol{X}}$ can be used instead of $\boldsymbol{X}$ in (8.56) and (8.58). On the other hand, we are unable to simply merge the similarity matrices, and instead we need to prepare S terms shown in (8.57).

Considering the above two points and also L^2 norm constraint over $\boldsymbol{\beta}$, we can then write the Lagrangian function as follows:

$$\begin{aligned} L(\boldsymbol{z}, \boldsymbol{\beta}) &= (\boldsymbol{z} - \boldsymbol{y})^\top (\boldsymbol{z} - \boldsymbol{y}) + \lambda_v (\boldsymbol{z} - \tilde{\boldsymbol{X}}\boldsymbol{\beta})^\top (\boldsymbol{z} - \tilde{\boldsymbol{X}}\boldsymbol{\beta}) \\ &+ \lambda_b \boldsymbol{\beta}^\top \boldsymbol{\beta} + \sum_s^S \lambda_g^{(s)} \boldsymbol{z}^\top \boldsymbol{L}^{(s)} \boldsymbol{z}, \end{aligned} \tag{8.62}$$

where $\boldsymbol{L}^{(s)}$ is graph Laplacian of the s-th similarity matrix (adjacency matrix or graph).

We set the partial derivative of this Lagrangian with respect to each parameter to zero, and obtain the following update rules:

$$\frac{\partial L(\boldsymbol{z}, \boldsymbol{\beta})}{\partial \boldsymbol{z}} = 0 \quad \Rightarrow \quad \hat{\boldsymbol{z}} = ((1 + \lambda_v)\boldsymbol{I} + \sum_s \lambda_g^{(s)} \boldsymbol{L}^{(s)})^{-1} (\boldsymbol{y} + \lambda_v \tilde{\boldsymbol{X}} \boldsymbol{\beta}). \tag{8.63}$$

$$\frac{\partial L(\boldsymbol{z}, \boldsymbol{\beta})}{\partial \boldsymbol{\beta}} = 0 \quad \Rightarrow \quad \hat{\boldsymbol{\beta}} = (\tilde{\boldsymbol{X}}^\top \tilde{\boldsymbol{X}} + \lambda_\beta \boldsymbol{I})^{-1} \tilde{\boldsymbol{X}}^\top \boldsymbol{z}, \tag{8.64}$$

Procedure 8.6 shows a pseudocode of the above procedure of estimating two parameter vectors $\boldsymbol{z}$ and $\boldsymbol{\beta}$, alternately. Note that if $R = 1$ and $S = 1$, this general case is equivalent to the simplest case of given $\boldsymbol{X}$ and $\boldsymbol{S}$.

8.2.2 Unsupervised Learning: Clustering

Two Matrices with Similarity Matrix, Sharing Instances

Similar to the supervised learning, this situation is the same as the unsupervised setting in Section 5.2.3 which integrates two types of data, i.e. a vector and a node in a graph, in the frame work of semisupervised clustering. We first briefly review this approach.

Let $\boldsymbol{Z}$ be cluster assignment matrix, which is used in many settings of clustering. We can directly use the ideas of constrained K-means clustering and

Procedure 8.6: MULTIPLE DATASETS INTEGRATIVE LINEAR REGRESSION.

Input: Matrices: $\boldsymbol{X}^{(1)}, \ldots, \boldsymbol{X}^{(R)}$; similarity matrices: $\boldsymbol{S}^{(1)}, \ldots, \boldsymbol{S}^{(S)}$
Output: Label assignment vector: $\boldsymbol{z}$; linear regression coefficient: $\boldsymbol{\beta}$

```
Compute X̃ from X^(1), ..., X^(R).;
foreach s do
    Compute graph Laplacian L^(s) from S^(s).;
Initialize z and β.;
repeat
    Update label assignment vector z by (8.63), fixing β.;
    Update weight vector β by (8.64), fixing z.;
until convergence;
Output estimated ẑ and β̂.;
```

spectral clustering for a vector and a node in a network, respectively, leading to the following two objective functions:

$$\min_{\boldsymbol{Z}} \operatorname{trace}(\boldsymbol{Z}^\top \boldsymbol{E} \boldsymbol{Z}), \qquad \min_{\boldsymbol{Z}} \operatorname{trace}(\boldsymbol{Z}^\top \boldsymbol{L} \boldsymbol{Z}), \tag{8.65}$$

respectively, where the (i, j)-element of $\boldsymbol{E}$, i.e. $\boldsymbol{E}_{ij}$, is

$$\boldsymbol{E}_{ij} = \frac{1}{2N} ||\boldsymbol{x}_i - \boldsymbol{x}_j||^2, \tag{8.66}$$

and $\boldsymbol{L}$ is graph Laplacian from the given adjacency (similarity) matrix.

Thus the Lagrangian can be written as follows:

$$L(\boldsymbol{Z}) = \operatorname{trace}(\boldsymbol{Z}^\top \boldsymbol{E} \boldsymbol{Z}) + \lambda_G \operatorname{trace}(\boldsymbol{Z}^\top \boldsymbol{L} \boldsymbol{Z}) - \lambda_C \operatorname{trace}(\boldsymbol{Z}^\top \boldsymbol{Z} - \boldsymbol{D}), \tag{8.67}$$

where λ_G and λ_C are hyperparameters. By setting the partial derivative of the Lagrangian with respect to $\boldsymbol{Z}$ to zero, we can see that $\boldsymbol{Z}$ can be estimated by first solving the following eigenvalue problem:

$$(\boldsymbol{E} + \lambda_G \boldsymbol{L}) \boldsymbol{Z} = \lambda_C \boldsymbol{Z} \tag{8.68}$$

and then after postprocessing, we can have the estimated $\hat{\boldsymbol{Z}}$. This is a possible approach for estimating cluster assignment matrix $\boldsymbol{Z}$, while both hyperparameters must be searched in optimization, in general, empirically.

Instead another approach is to introduce weights over the two data inputs and optimize them in the framework of a biconvex problem, which has two types of parameters, i.e. cluster assignment matrix $\boldsymbol{Z}$ and weights $\boldsymbol{w}$. Since we use weights over matrices, as mentioned in Section 6.2.1, we can conduct "weight learning", for which we have to change the above problem formulation into a maximization problem, in which we compute the similarity matrix $\boldsymbol{S}^{\boldsymbol{X}}$ for vectors, by using the simple (simplest) cosine similarity, as follows:

$$\boldsymbol{S}^{\boldsymbol{X}} = \boldsymbol{X} \boldsymbol{X}^\top, \tag{8.69}$$

and we can use the original input similarity matrix $\boldsymbol{S}$ for nodes.

Then $\text{trace}(\boldsymbol{Z}^\top \boldsymbol{S}^{\boldsymbol{X}} \boldsymbol{Z})$ is always nonzero, because of **Proposition 6.2**, and also all elements of $\boldsymbol{S}$ and $\boldsymbol{Z}$ are nonzero, and the following two traces are always nonzero:

$$\text{trace}(\boldsymbol{Z}^\top \boldsymbol{S}^{\boldsymbol{X}} \boldsymbol{Z}) \geq 0, \quad \text{trace}(\boldsymbol{Z}^\top \boldsymbol{S} \boldsymbol{Z}) \geq 0. \tag{8.70}$$

Then we can formulate a maximization problem with weights $\boldsymbol{w} = (w_{\boldsymbol{X}}, w_{\boldsymbol{S}})^\top$ ($w_{\boldsymbol{X}}$ and $w_{\boldsymbol{S}}$ for vectors and the similarity matrix, respectively) as follows:

$$\begin{aligned} \max_{\boldsymbol{w},\boldsymbol{Z}} \quad & w_{\boldsymbol{X}}\text{trace}(\boldsymbol{Z}^\top \boldsymbol{S}^{\boldsymbol{X}} \boldsymbol{Z}) + w_{\boldsymbol{S}}\text{trace}(\boldsymbol{Z}^\top \boldsymbol{S} \boldsymbol{Z}) \\ \text{subject to} \quad & \boldsymbol{Z}^\top \boldsymbol{Z} = \boldsymbol{D}, \ \ ||\boldsymbol{w}||^2 = \text{Const.} \end{aligned} \tag{8.71}$$

In fact this is the same formulation as a general case with two input matrices both have a feature set (no similarity matrices), explained in Section 8.1.2. Thus we can just follow the general case for estimating both parameters $\boldsymbol{w}$ and $\boldsymbol{Z}$ (see Section 8.1.2).

Generalization to R Matrices and S Similarity Matrices, Sharing Instances

First we can simply extend semisupervised learning for the general case, and the Lagrangian can be given as follows:

$$\begin{aligned} L(\boldsymbol{Z}) &= \sum_r^R \text{trace}(\boldsymbol{Z}^\top \boldsymbol{E}^{(r)} \boldsymbol{Z}) \\ &+ \lambda_G \sum_s^S \text{trace}(\boldsymbol{Z}^\top \boldsymbol{L}^{(s)} \boldsymbol{Z}) - \lambda_C \text{trace}(\boldsymbol{Z}^\top \boldsymbol{Z} - \boldsymbol{D}), \end{aligned} \tag{8.72}$$

where λ_G and λ_C are hyperparameters, and $\boldsymbol{E}^{(r)}$ and $\boldsymbol{L}^{(s)}$ are the extensions for the r-th and s-th inputs from $\boldsymbol{E}$ and $\boldsymbol{L}$ of the input of only two matrices including a similarity matrix.

Simply $\boldsymbol{Z}$ can be estimated by solving the following eigenvalue problem:

$$\left(\sum_r \boldsymbol{E}^{(r)} + \lambda_G \sum_s \boldsymbol{L}^{(s)}\right)\boldsymbol{Z} = \lambda_C \boldsymbol{Z}, \tag{8.73}$$

and running the postprocessing step over the obtained $\boldsymbol{Z}$.

On the other hand, if we try to avoid estimating hyperparameters empirically, as we have shown in the case of only two input matrices, we can conduct "weight learning", originally introduced in Section 6.2.1. That is, the formulation can be totally the same as that introduced for given R matrices sharing the same instance set, in Section 8.1.2. Thus we can just follow the above case with given R matrices, just by changing a half set of matrices to a set of similarity matrices.

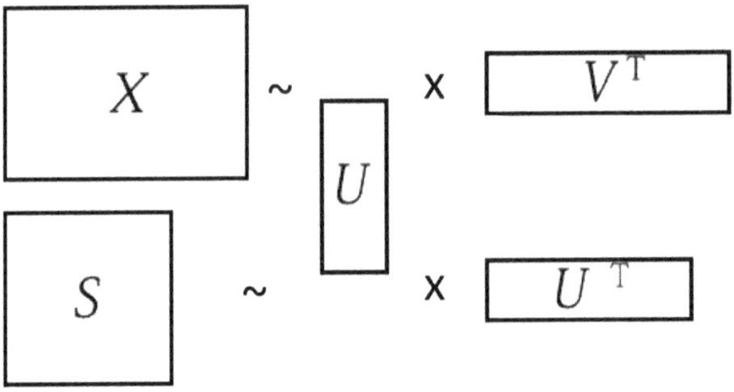

Figure 8.9: Collaborative matrix factorization, which factorizes given two matrices into two low-rank matrices: one is shared between the input two matrices and the other one is smooth over the similarity matrix.

8.2.3 Unsupervised Learning: Collaborative Matrix Factorization

Two Matrices with Similarity Matrix, Sharing Instances

We first consider only two inputs, i.e. $\boldsymbol{X}$ and $\boldsymbol{S}$. As mentioned already, a baseline approach is to generate two low-rank matrices by decomposing each given input matrix, but again clearly this way does not take advantage of the assumption of $\boldsymbol{X}$ and $\boldsymbol{S}$, which share the same set of instances and also $\boldsymbol{S}$ is a similarity (symmetric) matrix.

The two input matrices share the set of instances, and so the resultant low-rank matrices generated by factorization should be shared, as done in Section 8.1.3. Then let $\boldsymbol{U}$ be the shared low-rank matrix, and let $\boldsymbol{V}$ be another low-rank matrix, which is used to reproduce $\boldsymbol{X}$, as follows:

$$\boldsymbol{X} \sim \boldsymbol{U}\boldsymbol{V}^{\top} \tag{8.74}$$

Furthermore, $\boldsymbol{S}$ is a symmetric matrix, which means the two low-rank matrices from $\boldsymbol{S}$ must be the same matrix, i.e. $\boldsymbol{U}$, and this low-rank matrix, i.e. $\boldsymbol{U}$, should be smooth over $\boldsymbol{S}$ as follows:

$$\min_{\boldsymbol{U}} \mathrm{trace}(\boldsymbol{U}^{\top}\boldsymbol{L}\boldsymbol{U}), \tag{8.75}$$

where $\boldsymbol{L}$ $(= \boldsymbol{D} - \boldsymbol{S})$ is graph Laplacian for $\boldsymbol{S}$ and i-th diagonal element d_{ii} of diagonal matrix $\boldsymbol{D}$ is $d_{ii} = \sum_j \boldsymbol{S}_{ij}$ and $\boldsymbol{S}_{ij}$ is the (i,j)-th element of $\boldsymbol{S}$.

Fig. 8.9 shows a schematic picture of collaborative matrix factorization of this case. Overall we can formulate this problem, considering L^2 norm constraints, as follows:

$$\begin{aligned} \min_{\boldsymbol{U},\boldsymbol{V}} \quad & ||\boldsymbol{X} - \boldsymbol{U}\boldsymbol{V}^{\top}||^2 \\ \text{subject to} \quad & \mathrm{trace}(\boldsymbol{U}^{\top}\boldsymbol{L}\boldsymbol{U}) < \text{Const.},\ ||\boldsymbol{U}||^2 = \text{Const.},\ ||\boldsymbol{V}||^2 = \text{Const.} \end{aligned} \tag{8.76}$$

Procedure 8.7: COLLABORATIVE MATRIX FACTORIZATION FOR TWO MATRICES SHARING INSTANCES AND ONE MATRIX IS A SIMILARITY MATRIX.

Input: Matrix: $\boldsymbol{X}$, Similarity matrix $\boldsymbol{S}$; low-rank: K
Output: Low-rank matrices: $\boldsymbol{U}$, $\boldsymbol{V}$

```
1 Initialization: Initialize U and V.;
2 repeat
3     Update U by (8.77).;
4     Update V by (8.78).;
5 until convergence;
6 Output finally estimated Û and V̂.;
```

This is a biconvex problem with two matrix parameters, $\boldsymbol{U}$ and $\boldsymbol{V}$, which are estimated by repeating the following two steps alternately, which are 1) to update $\boldsymbol{U}$, fixing $\boldsymbol{V}$ and 2) to update $\boldsymbol{V}$, fixing $\boldsymbol{U}$.

We can then set up the Lagrangian function to be optimized from the above formulation, and by further setting the partial derivative of the Lagrangian with respect to each parameter to zero, we can see the update rules of the parameters. We can just show the result of multiplicative update rules, as follows:

$$U_{ik} \leftarrow U_{ik} \frac{(\boldsymbol{XV} + \lambda_U \boldsymbol{SU})_{ik}}{(\boldsymbol{UV}^\top \boldsymbol{V} + \lambda_U \boldsymbol{DU} + \lambda \boldsymbol{U})_{ik}}, \tag{8.77}$$

$$V_{jk} \leftarrow V_{jk} \frac{(\boldsymbol{X}^\top \boldsymbol{U})_{jk}}{(\boldsymbol{V}(\boldsymbol{U}^\top \boldsymbol{U} + \lambda \boldsymbol{I}))_{jk}}, \tag{8.78}$$

We can see that the update rule of $\boldsymbol{V}$ is totally the same as the update rule of regular matrix factorization, e.g. nonnegative matrix factorization. On the other hand, the update rule of $\boldsymbol{U}$ uses all information, including the input $\boldsymbol{X}$ and $\boldsymbol{S}$ and low rank matrices $\boldsymbol{U}$ and $\boldsymbol{V}$. **Procedure 8.7** shows a pseudocode of the procedure we have explained for collaborative matrix factorization of two input matrices sharing the same set of instances and one matrix is a similarity matrix.

Generalization to R Matrices and S Similarity Matrices, Sharing Instances

We then consider a general situation, in which we have an arbitrary number, say R, of regular matrices, $\boldsymbol{X}^{(1)}, \ldots, \boldsymbol{X}^{(R)}$, and also an arbitrary number, say S, of similarity matrices, $\boldsymbol{S}^{(1)}, \ldots, \boldsymbol{S}^{(S)}$. As we have shown collaborative matrix factorization in Section 8.1.3, we can consider low-rank matrix $\boldsymbol{V}^{(r)}$ for each input $\boldsymbol{X}^{(r)}$, as follows:

$$\boldsymbol{X}^{(r)} \sim \boldsymbol{U}(\boldsymbol{V}^{(r)})^\top (r = 1, \ldots, R) \tag{8.79}$$

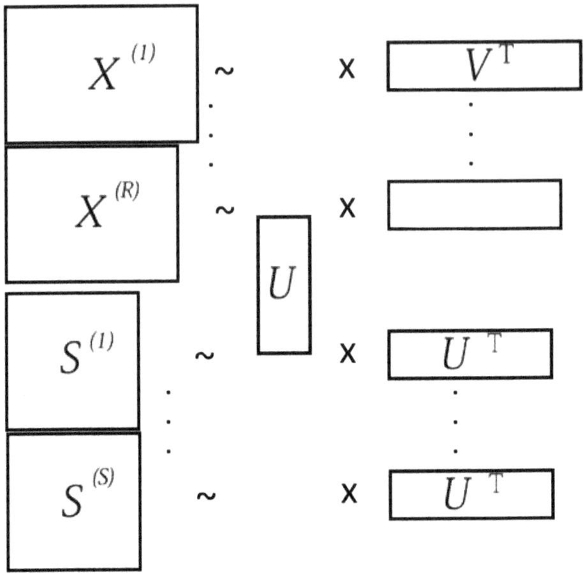

Figure 8.10: Collaborative matrix factorization, which factorizes given $(R+S)$ matrices into $R+1$ matrices: one is shared by all matrices and each of other R matrices is specific to one of the inputs $\boldsymbol{X}^{(r)}(r=1,\ldots,R)$.

However, $\boldsymbol{U}$ is shared by the both two sides, i.e. both the $\boldsymbol{X}$ and $\boldsymbol{S}$ sides, by which we cannot change $\boldsymbol{U}$, depending on $\boldsymbol{S}^{(s)}(s=1,\ldots,S)$. Also each of $\boldsymbol{S}^{(s)}(s=1,\ldots,S)$ is symmetric, meaning that each of them should be smoothed by $\boldsymbol{U}$.

$$\min_{\boldsymbol{U}} \sum_s \mathrm{trace}(\boldsymbol{U}^\top \boldsymbol{L}^{(s)} \boldsymbol{U}), \tag{8.80}$$

where $\boldsymbol{L}^{(s)}(=\boldsymbol{D}^{(s)}-\boldsymbol{S}^{(s)})$ is graph Laplacian for $\boldsymbol{S}^{(s)}$ and i-th diagonal element $d_{ii}^{(s)}$ of diagonal matrix $\boldsymbol{D}^{(s)}$ is $d_{ii}^{(s)}=\sum_j \boldsymbol{S}_{ij}^{(s)}$ and $\boldsymbol{S}_{ij}^{(s)}$ is the (i,j)-th element of $\boldsymbol{S}^{(s)}$.

Fig. 8.10 shows a schematic picture of collaborative matrix factorization, summarizing the above ideas. Then we can formulate our problem, considering L^2 norm constraints, as follows:

$$\begin{aligned}
\min_{\boldsymbol{U},\boldsymbol{V}^{(1)},\ldots,\boldsymbol{V}^{(R)}} \quad & \sum_r ||\boldsymbol{X}-\boldsymbol{U}(\boldsymbol{V}^{(r)})^\top||^2 \\
\text{subject to} \quad & \mathrm{trace}(\boldsymbol{U}^\top \boldsymbol{L}^{(s)} \boldsymbol{U}) < \text{Const.} \;\; (s=1,\ldots,S), \\
& ||\boldsymbol{U}||^2 = \text{Const.}, \;\; ||\boldsymbol{V}^{(r)}||^2 = \text{Const.} \;\; (r=1,\ldots,R).
\end{aligned} \tag{8.81}$$

To estimate a local minimum, we can take an alternate approach, which repeats

Procedure 8.8: COLLABORATIVE MATRIX FACTORIZATION FOR $R+S$ MATRICES SHARING INSTANCES AND ALSO S MATRICES ARE SIMILARITY MATRICES.

Input: Matrices: $\boldsymbol{X}^{(r)}(r=1,\ldots,R)$, Similarity matrices $\boldsymbol{S}^{(s)}(s=1,\ldots,S)$; low-rank: K
Output: Low-rank matrices: $\boldsymbol{U}$, $\boldsymbol{V}^{(r)}$ $(r=1,\ldots,R)$

```
Initialization: Initialize U and V^(r) (r = 1,...,R).;
repeat
    Update U by (8.82).;
    Update V^(r)(r = 1,...,R) by (8.83).;
until convergence;
Output finally estimated Û and V^(r)hat (r = 1,...,R).;
```

estimating one parameter, fixing all other parameters, for each parameter alternately.

From the above formulation, we can derive multiplicative update rules of each matrix parameter as follows:

$$U_{ik} \leftarrow U_{ik}\frac{\sum_r(\boldsymbol{X}^{(r)}\boldsymbol{V}^{(r)}+\lambda_U\sum_s \boldsymbol{S}^{(s)}\boldsymbol{U})_{ik}}{(\boldsymbol{U}\sum_r(\boldsymbol{V}^{(r)})^{\top}\boldsymbol{V}^{(r)}+\lambda_U\sum_s \boldsymbol{D}^{(s)}\boldsymbol{U}+\lambda \boldsymbol{U})_{ik}}, \tag{8.82}$$

$$V^{(r)}_{jk} \leftarrow V^{(r)}_{jk}\frac{((\boldsymbol{X}^{(r)})^{\top}\boldsymbol{U})_{jk}}{(\boldsymbol{V}^{(r)}(\boldsymbol{U}^{\top}\boldsymbol{U}+\gamma_r\boldsymbol{I}))_{jk}} \quad (r=1,\ldots,R), \tag{8.83}$$

where λ_U and $\gamma_r(r=1,\ldots,R)$ are hyperparameters.

From the multiplicative update rules, we can see the same properties as those in the case of two inputs, where one is a similarity matrix. That is, the update rules of $\boldsymbol{V}^{(r)}$ $(r=1,\ldots,R)$ are still the same as the case in which just a single matrix, say $\boldsymbol{X}$, is factorized into $\boldsymbol{U}$ and $\boldsymbol{V}^{\top}$. On the other hand, again the update rule of $\boldsymbol{V}^{(r)}$ depends upon all other information, including all inputs and all other matrix parameters. **Procedure 8.8** shows a pseudocode of the above procedure of estimating $R+1$ matrix parameters, in an alternate manner.

8.3 Matrices Sharing Instances and Features

Now we consider the input of three matrices, which share instances and features, as follows: Let $\boldsymbol{A}$, $\boldsymbol{B}$ and $\boldsymbol{C}$ be three matrices, where $\boldsymbol{A}$ and $\boldsymbol{B}$ share instances and $\boldsymbol{A}$ and $\boldsymbol{C}$ share features. Fig. 8.11 shows a schematic picture of the three matrices, $\boldsymbol{A}$, $\boldsymbol{B}$ and $\boldsymbol{C}$.

The difference of this input from those in Sections 8.1 and 8.2 is $\boldsymbol{C}$, which shares features with $\boldsymbol{A}$, meaning that $\boldsymbol{C}$ has nothing to do with labels of $\boldsymbol{A}$, even if instances of $\boldsymbol{A}$ has labels. We can then say that if instances of $\boldsymbol{A}$ have labels, supervised learning with $\boldsymbol{A}$ and $\boldsymbol{B}$ were already considered in Sections 8.1.1 and 8.2.1. Thus we do not consider any supervised learning setting here. Also we

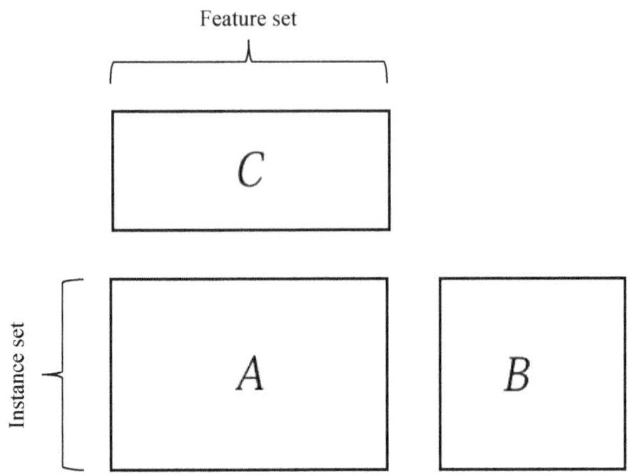

Figure 8.11: A schematic picture of the input: three matrices (**A**, **B** and **C**), sharing in instances by **A** and **B** and in features by **A** and **C**.

have two feature sets in these three matrices, i.e. **A** and **B** (features of **C** are the same as **A**). These two matrices share the same set of instances, and this setting was also already considered in Section 8.1.4. Thus we do not consider approaches for examining interactions between features of two sets, **A** and **B**. Thus for this input, we consider two machine learning paradigms, i.e. unsupervised learning of clustering and (collaborative) matrix factorization. Below under each of these two machine learning paradigms, we discuss the approaches for the input of three matrices sharing both instances and features, and then a more general situation in which we have R matrices, which share instances with S matrices and also share features with T matrices.

In fact this input is very important in real world. A typical real-world example is the input of content-based filtering, in which **A** is a user-item matrix, for which we can have user demographic data (**B**) and item contents information (**C**). For example if items are movies, we can have contents information on movie topics (like drama, action and thriller, etc.), directors, actors and actresses, etc. of each movie. If the items are foods, you can have information for ingredients, the produced locations (like a country and region), and dates.

As a special case, **B**, **C** or both **B** and **C**, can be similarity matrices, i.e. **B** for user similarities and **C** for item similarities. Similarities might be more easily given and would be useful.

Also we can have multiple matrices for **A**, like that data can be retrieved from multiple E-commerce sites or data can be measured in a time-series manner. Also for **B** and **C** we might have multiple matrices for each of them, because data might be gathered from multiple different resources.

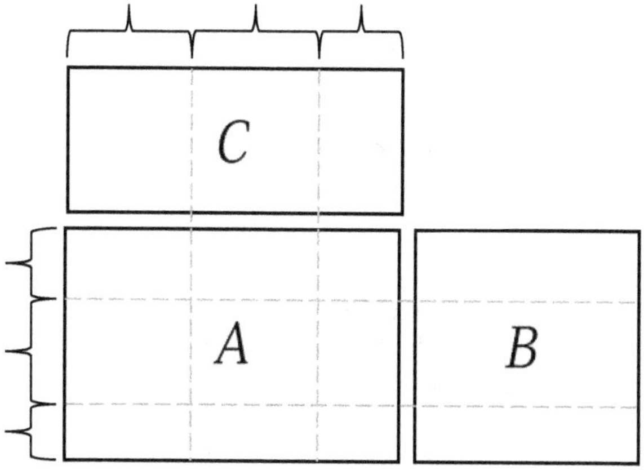

Figure 8.12: A schematic picture of biclustering three input matrices ($\boldsymbol{A}$, $\boldsymbol{B}$ and $\boldsymbol{C}$), where the two sides are clustered independently.

8.3.1 Unsupervised Learning: Clustering

Three Matrices, Sharing Instances (between Two Matrices) and Features (between Two Matrices)

The input matrices share both instances and features, and there are following two points, being related with this setting:

- We examined clustering for two matrices, which share instances, in Section 8.1.2.
- In Section 7.1.1, for a single matrix, we showed biclustering, in which both two sides, i.e. instances and features are clustered independently.

Thus we can combine these two approaches, leading to biclustering of our input of three matrices, sharing instances and features. Fig. 8.12 shows a schematic picture of biclustering over the input three matrices, $\boldsymbol{A}$, $\boldsymbol{B}$ and $\boldsymbol{C}$.

In Section 8.1.2, we formulated the clustering problem as a maximization problem, i.e. weight learning. This formulation is possible here also. However we can keep a minimization problem of clustering, by keeping using hyperparameters, because this is a simple manner and we do not use so many hyperparameters in this setting. In fact setting up the minimization problem was already briefly given in Section 8.1.2 for two matrices $\boldsymbol{X}$ and $\boldsymbol{Y}$, which share instances. That is, first the objective functions of clustering for $\boldsymbol{A}$ and $\boldsymbol{B}$ can be given independently as follows:

$$\min_{\boldsymbol{Z}} \operatorname{trace}(\boldsymbol{Z}^\top \boldsymbol{E}^A \boldsymbol{Z}), \qquad \min_{\boldsymbol{Z}} \operatorname{trace}(\boldsymbol{Z}^\top \boldsymbol{E}^B \boldsymbol{Z}), \tag{8.84}$$

respectively, where the (i,j) elements of $\boldsymbol{E}^A$ and $\boldsymbol{E}^B$ are

$$\boldsymbol{E}^A_{ij} = ||\boldsymbol{x}^A_i - \boldsymbol{x}^A_j||^2, \quad \boldsymbol{E}^B_{ij} = ||\boldsymbol{x}^B_i - \boldsymbol{x}^B_j||^2, \tag{8.85}$$

respectively, where $\boldsymbol{x}_i^A$ is the i-th instance of $\boldsymbol{A}$.

Then we can simply formulate the clustering problem for two matrices, $\boldsymbol{A}$ and $\boldsymbol{B}$, of our input, as follows:

$$\min_{\boldsymbol{Z}} \quad \mathrm{trace}(\boldsymbol{Z}^\top \boldsymbol{E}^A \boldsymbol{Z})$$
$$\text{subject to} \quad \mathrm{trace}(\boldsymbol{Z}^\top \boldsymbol{E}^B \boldsymbol{Z}) < \text{Const.} \quad \boldsymbol{Z}^\top \boldsymbol{Z} = \boldsymbol{D}, \tag{8.86}$$

where $\boldsymbol{D}$ is a diagonal matrix.

We can then generate the Lagrangian function from this formulation and by setting the partial derivative of the Lagrangian function with respect to $\boldsymbol{Z}$ to zero, we can have the following eigenvalue problem.

$$(\boldsymbol{E}^A + \lambda_B \boldsymbol{E}^B)\boldsymbol{Z} = \lambda \boldsymbol{Z} \tag{8.87}$$

This is equivalent to clustering of learning multiple data types (data integrative learning) in Section 5.2.3. In particular, if $\boldsymbol{B}$ is a similarity matrix, $\boldsymbol{B}$ can be regarded as a graph or an adjacency matrix. Then (8.87) can be the same as (5.79) in Section 5.2.3, where given a matrix and a graph, a vector in the matrix and a node in the graph were combined as an instance for clustering. Then by solving this eigenvalue problem of (8.87) and running a postprocessing procedure of the resultant $\boldsymbol{Z}$, we can have the finally estimated $\hat{\boldsymbol{Z}}$, which allows to show clusters over instances.

Similarly we can do clustering over features. Let $\boldsymbol{H}$ be the cluster assignment matrix for the feature side, which is equivalent to $\boldsymbol{Z}$ of the instance side. We can then formulate the clustering problem for the features side in a totally similar way to the instance side, as follows:

$$\min_{\boldsymbol{H}} \quad \mathrm{trace}(\boldsymbol{H}^\top \boldsymbol{F}^A \boldsymbol{H})$$
$$\text{subject to} \quad \mathrm{trace}(\boldsymbol{H}^\top \boldsymbol{F}^C \boldsymbol{H}) < \text{Const.}, \quad \boldsymbol{H}^\top \boldsymbol{H} = \boldsymbol{D}, \tag{8.88}$$

where the (i, j) element of $\boldsymbol{F}^A$ and $\boldsymbol{F}^C$ are written as follows:

$$\boldsymbol{F}_{ij}^A = ||\boldsymbol{f}_i^A - \boldsymbol{f}_j^A||^2, \quad \boldsymbol{F}_{ij}^C = ||\boldsymbol{f}_i^C - \boldsymbol{f}_j^C||^2, \tag{8.89}$$

respectively, where $\boldsymbol{f}_i^A$ is the i-th feature vector (column) of $\boldsymbol{A}$.

This is equivalent to the formulation of the instance side, i.e. (8.86), and also the solution of this formulation results into the following eigenvalue problem:

$$(\boldsymbol{F}^A + \lambda_C \boldsymbol{F}^C)\boldsymbol{H} = \lambda \boldsymbol{H} \tag{8.90}$$

By solving this eigenvalue problem, and also running the postprocessing procedure over the obtained $\boldsymbol{H}$, we can have the estimated $\hat{\boldsymbol{H}}$, which allows to show clusters of features.

We then sort both instances and features of $\boldsymbol{A}$, independently, according to the clustering results, i.e. $\hat{\boldsymbol{Z}}$ and $\hat{\boldsymbol{H}}$, and then finally show the resultant $\boldsymbol{A}$ as a biclustering result. **Procedure 8.9** shows a pseudocode of the above biclustering procedure. Note that this procedure does not contain the process of deciding the

Procedure 8.9: BICLUSTERING FOR 3 MATRICES SHARING INSTANCES AND FEATURES.

Input: Matrices: $\boldsymbol{A}$, $\boldsymbol{B}$ and $\boldsymbol{C}$
Output: Cluster assignment matrices: instance side $\boldsymbol{Z}$; feature side: $\boldsymbol{H}$

Solve the eigenvalue problem given by (8.87) to have $\boldsymbol{Z}$.;
Run K-`means Clustering(`$\boldsymbol{Z}$`)` to have the finally estimated $\hat{\boldsymbol{Z}}$.;
Solve the eigenvalue problem given by (8.90) to have $\boldsymbol{H}$.;
Run K-`means Clustering(`$\boldsymbol{H}$`)` to have the finally estimated $\hat{\boldsymbol{H}}$.;
Sort instances in $\boldsymbol{A}$, according to $\hat{\boldsymbol{Z}}$.;
Sort features in $\boldsymbol{A}^\top$, according to $\hat{\boldsymbol{H}}$.;
Show sorted $\boldsymbol{A}$ on the 2D space.;

values of hyperparameters. Thus unless hyperparameters are given by some prior knowledge, we need to determine hyperparameters empirically.

Of course instead of the minimization problem, which is regularly used for clustering, we can use the maximization problem, which was shown in Section 8.1.2 and originally shown in Section 6.2.1. The maximization problem allows us to conduct "weight learning", which can decide the weights over the multiple matrices. However the input here has only two matrices for each of the instance and feature sides, and determining regularization coefficients (i.e. hyperparameters) empirically would not be computationally heavy. Thus we think the simple minimization solution would be good enough for this application.

R Matrices, Sharing Instances with S Matrices and Features with T Matrices

We then consider a general situation, which has R matrices, which share instances with other S matrices and also share features with other T matrices: Let $\boldsymbol{A}^{(1)},\ldots,\boldsymbol{A}^{(R)}$, $\boldsymbol{B}^{(1)},\ldots,\boldsymbol{B}^{(S)}$ and $\boldsymbol{C}^{(1)},\ldots,\boldsymbol{C}^{(T)}$, be matrices, where $\boldsymbol{A}^{(*)}$ and $\boldsymbol{B}^{(*)}$ share instances and $\boldsymbol{A}^{(*)}$ and $\boldsymbol{C}^{(*)}$ share features. Note that all $\boldsymbol{A}^{(*)}$ keep the same matrix size, i.e. $(N \times M)$-matrices. On the other hand, all $\boldsymbol{B}^{(*)}$ keep the same size of instances, but the size of features can differ. Also all $\boldsymbol{C}^{(*)}$ must have the same size of features but can have different sizes of instances for $\boldsymbol{C}^{(*)}$. Fig. 8.11 shows a schematic picture of this input.

Similar to that for the input of three matrices, if we formulate the problem as a minimization problem, we can have the resultant eigenvalue problem for instances, as follows:

$$(\sum_r \boldsymbol{E}^{A,(r)} + \lambda_B \sum_s \boldsymbol{E}^{B,(s)})\boldsymbol{Z} = \lambda \boldsymbol{Z}, \tag{8.91}$$

where λ_B is a hyperparameter (λ becomes eigenvalues), and the (i,j) elements of $\boldsymbol{E}^{A,(r)}$ and $\boldsymbol{E}^{B,(s)}$ can be given as follows:

$$\boldsymbol{E}_{ij}^{A,(r)} = ||\boldsymbol{x}_i^{A,(r)} - \boldsymbol{x}_j^{A,(r)}||^2, \quad \boldsymbol{E}_{ij}^{B,(s)} = ||\boldsymbol{x}_i^{B,(s)} - \boldsymbol{x}_j^{B,(s)}||^2, \tag{8.92}$$

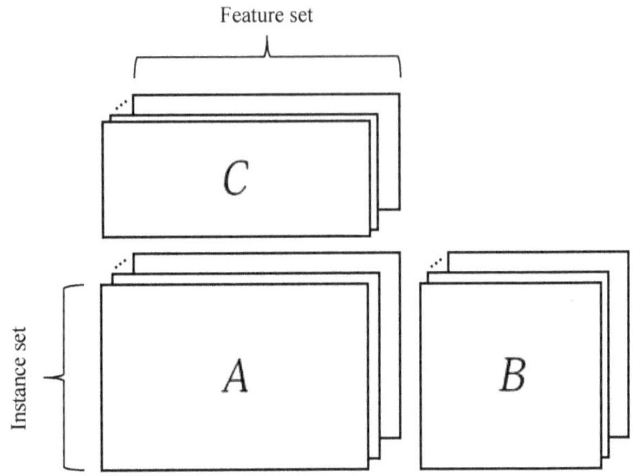

Figure 8.13: A schematic picture of the input of three types of matrices: $(\boldsymbol{A}^{(1)}, \ldots, \boldsymbol{A}^{(R)}, \boldsymbol{B}^{(1)}, \ldots, \boldsymbol{B}^{(S)}$ and $\boldsymbol{C}^{(1)}, \ldots, \boldsymbol{C}^{(T)})$, where $\boldsymbol{A}^{(*)}$ and $\boldsymbol{B}^{(*)}$ share instances, and $\boldsymbol{A}^{(*)}$ and $\boldsymbol{C}^{(*)}$ share features.

respectively, where $\boldsymbol{x}_i^{A,(r)}$ is the i-th instance of the r-th matrix among $\boldsymbol{A}^{(1)}, \ldots, \boldsymbol{A}^{(R)}$.

We can do clustering for the feature side as well. We then have the resultant eigenvalue problem for cluster assignment matrix $\boldsymbol{H}$ for features, as follows:

$$(\sum_r \boldsymbol{F}^{A,(r)} + \lambda_C \sum_t \boldsymbol{F}^{C,(t)})\boldsymbol{H} = \lambda \boldsymbol{H}, \tag{8.93}$$

where λ_C is a hyperparameter (λ can be obtained as eigenvalues), and the (i, j) elements of $\boldsymbol{F}^{A,(r)}$ and $\boldsymbol{F}^{C,(t)}$ are written as follows:

$$\boldsymbol{F}_{ij}^{A,(r)} = ||\boldsymbol{f}_i^{A,(r)} - \boldsymbol{f}_j^{A,(r)}||^2, \quad \boldsymbol{F}_{ij}^{C,(t)} = ||\boldsymbol{f}_i^{C,(t)} - \boldsymbol{f}_j^{C,(t)}||^2, \tag{8.94}$$

respectively, where $\boldsymbol{f}_i^A$ is the i-th feature vector (column) of $\boldsymbol{A}$.

Then by solving the above two eigenvalue problems and also running postprocessing procedures over the obtained cluster assignment matrices, we can have the estimated $\hat{\boldsymbol{Z}}$ and $\hat{\boldsymbol{H}}$, which show the resultant clusters of both sides. **Procedure 8.10** shows a pseudocode of the above biclustering procedure. As you can see from this procedure, even if the number of matrices increases, the procedure is fundamentally kept the same as that in **Procedure 8.9**.

Again the above minimization problem keeps hyperparameters, i.e. regularization coefficients, and if we use only three matrices, the number of hyperparameters is only two, which would not be a big problem to decide. However as the number of given matrices increases, empirically deciding the optimum value of each regularization coefficient is harder. Then for this issue, formulating the problem as the maximization problem, which can determine the weights over the matrices from data, might be one possible alternative. That is, we can conduct "weight learning", which was originally introduced in Section 6.2.1. More closely, Section 8.1.2

Procedure 8.10: BICLUSTERING FOR MULTIPLE MATRICES SHARING INSTANCES AND FEATURES.

Input: Matrices: $\boldsymbol{A}^{(r)}(r = 1, \ldots, R)$, $\boldsymbol{B}^{(s)}(s = 1, \ldots, S)$ and $\boldsymbol{C}^{(t)}(t = 1, \ldots, T)$

Output: Cluster assignment matrices: instance side $\boldsymbol{Z}$; feature side: $\boldsymbol{H}$

Solve the eigenvalue problem given by (8.91) to have $\boldsymbol{Z}$.;

Run K`-means Clustering(`$\boldsymbol{Z}$`)` to have the finally estimated $\hat{\boldsymbol{Z}}$.;

Solve the eigenvalue problem given by (8.93) to have $\boldsymbol{H}$.;

Run K`-means Clustering(`$\boldsymbol{H}$`)` to have the finally estimated $\hat{\boldsymbol{H}}$.;

Sort instances in $\boldsymbol{A}^{(*)}$, according to $\hat{\boldsymbol{Z}}$.;

Sort features in $(\boldsymbol{A}^{(*)})^\top$, according to $\hat{\boldsymbol{H}}$.;

Show sorted $\boldsymbol{A}^{(*)}$ on the 2D space.;

shows the procedure of clustering instances when two types of matrices sharing the instances, corresponding to $\boldsymbol{A}^{(*)}$ and $\boldsymbol{B}^{(*)}$. That is, **Procedure 8.1** corresponds to clustering instances from $\boldsymbol{A}^{(*)}$ and $\boldsymbol{B}^{(*)}$ to have cluster assignment matrix $\boldsymbol{Z}$. We can then run **Procedure 8.1** over features of $\boldsymbol{A}^{(*)}$ and $\boldsymbol{C}^{(*)}$ to obtain cluster assignment matrix $\boldsymbol{H}$ of the feature side. Then finally obtained cluster assignment matrices $\hat{\boldsymbol{Z}}$ and $\hat{\boldsymbol{H}}$ can be used for biclustering. Thus the entire procedure is the same as **Procedure 8.10**, while the clustering step for estimating $\boldsymbol{Z}$ and $\boldsymbol{H}$ can be replaced with **Procedure 8.1**, which allows to perform clustering by solving a maximization problem to estimate $\hat{\boldsymbol{Z}}$ (and $\hat{\boldsymbol{H}}$ by applying this procedure to the feature side).

8.3.2 Unsupervised Learning: Collaborative Matrix Factorization

Three Matrices, A, B and C, Sharing Instances (between A and B) and Features (between A and C)

In Section 8.1.3, we have seen a baseline of collaborative matrix factorization, which factorizes given two matrices $\boldsymbol{X}$ and $\boldsymbol{Y}$, sharing instances, into three low-rank matrices. Fig. 8.5 shows a schematic picture of the three low-rank matrices as $\boldsymbol{U}$, $\boldsymbol{V}$ and $\boldsymbol{W}$, where the point is that $\boldsymbol{U}$ is shared by the two factorization, meaning that $\boldsymbol{U}$ captures shared properties between $\boldsymbol{X}$ and $\boldsymbol{Y}$, while the other two low-rank matrices, $\boldsymbol{V}$ and $\boldsymbol{W}$, should show information specific to each of $\boldsymbol{X}$ and $\boldsymbol{Y}$, respectively.

We can now follow this idea and also apply to both the instance and feature sides of given matrices.

That is, $\boldsymbol{A}$ is factorized into two low-rank matrices, $\boldsymbol{U}$ and $\boldsymbol{V}$, and which are also low-rank matrices obtained by factorizing $\boldsymbol{B}$ and $\boldsymbol{C}$, respectively. Fig. 8.14 shows a schematic picture of this collaborative matrix factorization. As shown in this figure, for the instance side, $\boldsymbol{A}$ and $\boldsymbol{B}$ are factorized into $(\boldsymbol{U}, \boldsymbol{V})$ and $(\boldsymbol{U}, \boldsymbol{P})$,

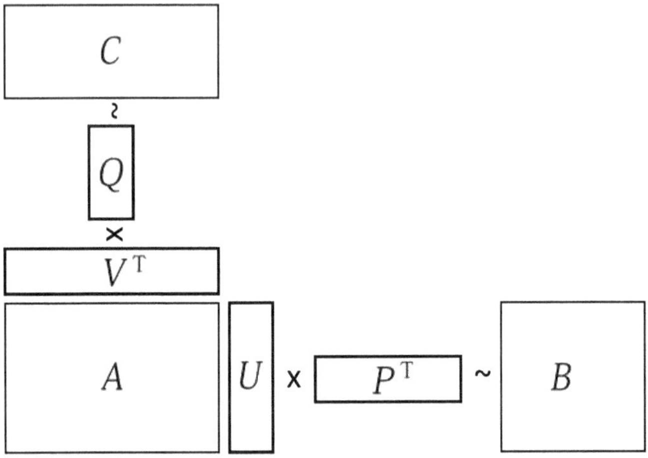

Figure 8.14: Collaborative matrix factorization, which factorizes given three matrices into four low-rank matrices.

respectively, sharing $\boldsymbol{U}$. Also for the feature side, $\boldsymbol{A}$ and $\boldsymbol{C}$ are factorized into ($\boldsymbol{U}$, $\boldsymbol{V}$) and ($\boldsymbol{V}$, $\boldsymbol{Q}$), respectively, sharing $\boldsymbol{V}$. Thus their approximations can be written as follows:

$$\boldsymbol{A} \sim \boldsymbol{U}\boldsymbol{V}^\top, \ \ \boldsymbol{B} \sim \boldsymbol{U}\boldsymbol{P}^\top, \ \ \boldsymbol{C} \sim \boldsymbol{Q}\boldsymbol{V}^\top \tag{8.95}$$

We can then formulate this collaborative matrix factorization, considering L^2 norm for constraints of the matrix parameters, as follows:

$$\begin{aligned} \min_{\boldsymbol{U},\boldsymbol{V},\boldsymbol{P},\boldsymbol{Q}} \quad & ||\boldsymbol{A} - \boldsymbol{U}\boldsymbol{V}^\top||^2 + \lambda_B||\boldsymbol{B} - \boldsymbol{U}\boldsymbol{P}^\top||^2 + \lambda_C||\boldsymbol{C} - \boldsymbol{Q}\boldsymbol{V}^\top||^2 \\ \text{subject to} \quad & ||\boldsymbol{U}||^2 = \text{Const.}, \ ||\boldsymbol{V}||^2 = \text{Const.}, \ ||\boldsymbol{P}||^2 = \text{Const.}, \ ||\boldsymbol{Q}||^2 = \text{Const.}, \end{aligned} \tag{8.96}$$

where λ_B and λ_C are hyperparameters.

We can then write the Lagrangian function, following the above formulation as follows:

$$\begin{aligned} L(\boldsymbol{U},\boldsymbol{V},\boldsymbol{P},\boldsymbol{Q}) &= ||\boldsymbol{A} - \boldsymbol{U}\boldsymbol{V}^\top||^2 + \lambda_B||\boldsymbol{B} - \boldsymbol{U}\boldsymbol{P}^\top||^2 + \lambda_C||\boldsymbol{C} - \boldsymbol{Q}\boldsymbol{V}^\top||^2 \\ &+ \lambda(||\boldsymbol{U}||^2 + ||\boldsymbol{V}||^2 + ||\boldsymbol{P}||^2 + ||\boldsymbol{Q}||^2), \end{aligned} \tag{8.97}$$

where λ is a hyperparameter.

Similar to the estimation of matrix parameters of other collaborative matrix factorization, we can repeat the following step for all parameters: estimate one parameter, fixing all other parameters. We can obtain the update rule for each parameter as the additive update rule by setting the partial derivative of the Lagrangian with respect to each parameter to zero.

First we can easily notice that for $\boldsymbol{U}$ ($\boldsymbol{V}$), by ignoring terms in (8.97) which are not related with $\boldsymbol{U}$ ($\boldsymbol{V}$), the Lagrangian function is equivalent to that of the input of two matrices sharing instances, i.e. (8.26):

$$\frac{\partial L(\boldsymbol{U},\boldsymbol{V},\boldsymbol{P},\boldsymbol{Q})}{\partial \boldsymbol{U}} = \boldsymbol{0} \;\Rightarrow\; \boldsymbol{U} \leftarrow (\boldsymbol{A}\boldsymbol{V} + \lambda_B \boldsymbol{B}\boldsymbol{P})(\boldsymbol{V}^\top\boldsymbol{V} + \lambda_B \boldsymbol{P}^\top\boldsymbol{P} + \lambda\boldsymbol{I})^{-1},$$
$$\frac{\partial L(\boldsymbol{U},\boldsymbol{V},\boldsymbol{P},\boldsymbol{Q})}{\partial \boldsymbol{V}} = \boldsymbol{0} \;\Rightarrow\; \boldsymbol{V}^\top \leftarrow (\boldsymbol{U}^\top\boldsymbol{U} + \lambda_C \boldsymbol{Q}^\top\boldsymbol{Q} + \lambda\boldsymbol{I})^{-1}(\boldsymbol{U}^\top\boldsymbol{A} + \lambda_C \boldsymbol{Q}^\top\boldsymbol{C}) \tag{8.98}$$

Second, for $\boldsymbol{P}$ ($\boldsymbol{Q}$), similarly by ignoring the terms in (8.97) which are not related with $\boldsymbol{P}$ ($\boldsymbol{Q}$), we can easily see that the Lagrangian function is equivalent to that of the input of just a single matrix, i.e. (7.21). Then we can derive the following additive update rules:

$$\frac{\partial L(\boldsymbol{U},\boldsymbol{V},\boldsymbol{P},\boldsymbol{Q})}{\partial \boldsymbol{P}} = \boldsymbol{0} \;\Rightarrow\; \boldsymbol{P}^\top \leftarrow (\boldsymbol{U}^\top\boldsymbol{U} + \lambda_P \boldsymbol{I})^{-1}\boldsymbol{U}^\top\boldsymbol{B}, \tag{8.99}$$
$$\frac{\partial L(\boldsymbol{U},\boldsymbol{V},\boldsymbol{P},\boldsymbol{Q})}{\partial \boldsymbol{Q}} = \boldsymbol{0} \;\Rightarrow\; \boldsymbol{Q} \leftarrow \boldsymbol{C}\boldsymbol{V}(\boldsymbol{V}^\top\boldsymbol{V} + \lambda_Q \boldsymbol{I})^{-1}, \tag{8.100}$$

where $\lambda_P = \frac{\lambda}{\lambda_B}$ and $\lambda_Q = \frac{\lambda}{\lambda_C}$.

Next due to the derivation of setting the partial derivatives with respect to parameters to zero, we can derive multiplicative rules as follows:

$$\boldsymbol{U}_{ik} \leftarrow \boldsymbol{U}_{ik}\frac{(\boldsymbol{A}\boldsymbol{V} + \lambda_B \boldsymbol{B}\boldsymbol{P})_{ik}}{(\boldsymbol{U}(\boldsymbol{V}^\top\boldsymbol{V} + \lambda_B \boldsymbol{P}^\top\boldsymbol{P} + \lambda\boldsymbol{I}))_{ik}}, \tag{8.101}$$
$$\boldsymbol{V}_{kj} \leftarrow \boldsymbol{V}_{kj}\frac{(\boldsymbol{A}^\top\boldsymbol{U} + \lambda_C \boldsymbol{C}^\top\boldsymbol{Q})_{ik}}{(\boldsymbol{V}(\boldsymbol{U}^\top\boldsymbol{U} + \lambda_C \boldsymbol{Q}^\top\boldsymbol{Q} + \lambda\boldsymbol{I}))_{kj}}, \tag{8.102}$$
$$\boldsymbol{P}_{jk} \leftarrow \boldsymbol{P}_{jk}\frac{(\boldsymbol{B}^\top\boldsymbol{U})_{jk}}{(\boldsymbol{P}(\boldsymbol{U}^\top\boldsymbol{U} + \lambda_P \boldsymbol{I}))_{jk}}, \tag{8.103}$$
$$\boldsymbol{Q}_{ik} \leftarrow \boldsymbol{Q}_{ik}\frac{(\boldsymbol{C}\boldsymbol{V})_{ik}}{(\boldsymbol{Q}(\boldsymbol{V}^\top\boldsymbol{V} + \lambda_Q \boldsymbol{I}))_{ik}}, \tag{8.104}$$

Procedure 8.11 shows a pseudocode of the above procedure for estimating four matrix parameters of given three matrices, which share both instances and features.

Three Matrices, A, B and C, Sharing Instances (between A and B) and Features (between C and C), where B and C are Similarity Matrices

If $\boldsymbol{B}$, $\boldsymbol{C}$ or both $\boldsymbol{B}$ and $\boldsymbol{C}$ are similarity matrices, i.e. adjacency matrices or graphs, we can make the entire computation in a difference way (and somewhat simpler) so that factorized low-rank matrices should be smoothed over the given similarity matrices. In fact in Section 8.2.3, we already have considered matrix factorization for the input of two matrices sharing instances and also one matrix being a similarity matrix. Thus here we consider the situation in which both

Procedure 8.11: COLLABORATIVE MATRIX FACTORIZATION FOR THREE MATRICES SHARING INSTANCES AND FEATURES.

Input: Three matrices: $\boldsymbol{A}$, $\boldsymbol{B}$ and $\boldsymbol{C}$; low-rank: K
Output: Low-rank matrices: $\boldsymbol{U}$, $\boldsymbol{V}$, $\boldsymbol{P}$ and $\boldsymbol{Q}$

```
Initialization: Initialize U, V, P and Q.;
repeat
    Update U by (8.101).;
    Update V by (8.102).;
    Update P by (8.103).;
    Update Q by (8.104).;
until convergence;
Output estimated Û, V̂, P̂ and Q̂.;
```

$\boldsymbol{B}$ and $\boldsymbol{C}$ are similarity matrices. A key point of this situation is that $\boldsymbol{P}$ and $\boldsymbol{Q}$ must be $\boldsymbol{U}$ and $\boldsymbol{V}$, because of symmetricity of similarity matrices. Fig. 8.15 is a schematic picture of showing input three matrices, $\boldsymbol{A}$, $\boldsymbol{B}$ and $\boldsymbol{C}$, and low-rank matrices $\boldsymbol{U}$ and $\boldsymbol{V}$, which are obtained by factorizing $\boldsymbol{A}$, $\boldsymbol{B}$ and $\boldsymbol{C}$. This figure clearly shows the relationships among $\boldsymbol{A}$, $\boldsymbol{B}$ and $\boldsymbol{C}$. That is, $\boldsymbol{B}$ and $\boldsymbol{C}$ are factorized into low-rank matrices $\boldsymbol{U}$ and $\boldsymbol{V}$, respectively, which are obtained by factorizing $\boldsymbol{A}$ as well.

Thus, the following approximation should be considered:

$$\boldsymbol{A} \sim \boldsymbol{U}\boldsymbol{V}^\top, \quad \boldsymbol{B} \sim \boldsymbol{U}\boldsymbol{U}^\top, \quad \boldsymbol{C} \sim \boldsymbol{V}\boldsymbol{V}^\top \tag{8.105}$$

We can then formulate this collaborative matrix factorization by using graph Laplacian of the similarity matrices, considering L^2 norm for constraints of the matrix parameters, as follows:

$$\begin{aligned} \min_{\boldsymbol{U},\boldsymbol{V}} \quad & ||\boldsymbol{A} - \boldsymbol{U}\boldsymbol{V}^\top||^2 + \lambda_B \mathrm{trace}(\boldsymbol{U}^\top \boldsymbol{L}^{(B)} \boldsymbol{U}) + \lambda_C \mathrm{trace}(\boldsymbol{V}^\top \boldsymbol{L}^{(C)} \boldsymbol{V}) \\ \text{subject to} \quad & ||\boldsymbol{U}||^2 = \text{Const.}, \ ||\boldsymbol{V}||^2 = \text{Const.}, \end{aligned} \tag{8.106}$$

where $\boldsymbol{L}^{(B)} (= \boldsymbol{D}^{(B)} - \boldsymbol{B})$ and $\boldsymbol{L}^{(C)} (= \boldsymbol{D}^{(C)} - \boldsymbol{C})$ are graph Laplacian from similarity (adjacency) matrices $\boldsymbol{B}$ and $\boldsymbol{C}$, respectively, and $\boldsymbol{D}^{(B)}$ and $\boldsymbol{D}^{(C)}$ are diagonal matrices, in which the k-th diagonal elements $d_k^{(B)}$ and $d_k^{(C)}$ are computed as $d_k^{(B)} = \sum_i \boldsymbol{B}_{ki}$ and $d_k^{(C)} = \sum_i \boldsymbol{C}_{ki}$, respectively, and λ_B and λ_C are hyperparameters.

In this formulation, the first term shows that $\boldsymbol{U}$ and $\boldsymbol{V}$ are low-rank matrices obtained by factorization of $\boldsymbol{A}$, and the second and third terms show that low-rank matrices $\boldsymbol{U}$ and $\boldsymbol{V}$ have to be smooth over $\boldsymbol{B}$ and $\boldsymbol{C}$, respectively.

We have only two matrix parameters, and the above problem is biconvex, meaning that we can reach the local minimum, a solution of this problem, by alternately repeating 1) estimating $\boldsymbol{U}$, fixing $\boldsymbol{V}$ and 2) estimating $\boldsymbol{V}$, fixing $\boldsymbol{U}$, until convergence.

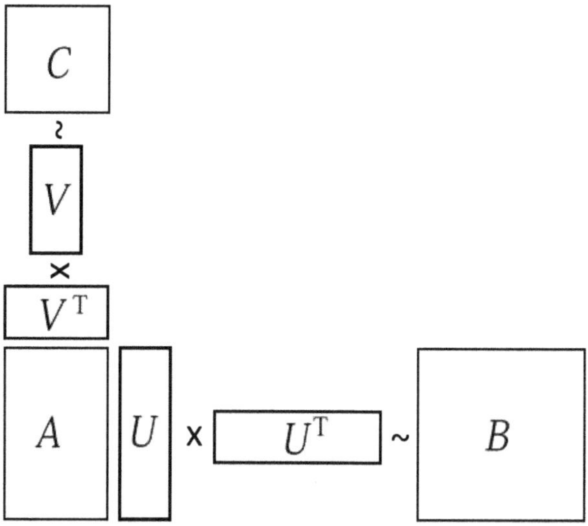

Figure 8.15: Collaborative matrix factorization, which factorizes given three matrices (including two similarity matrices) into two low-rank matrices.

We can then write the Lagrangian function, following the above formulation as follows:

$$\begin{aligned} L(\boldsymbol{U},\boldsymbol{V}) &= \|\boldsymbol{A}-\boldsymbol{U}\boldsymbol{V}^\top\|^2 + \lambda_B \mathrm{trace}(\boldsymbol{U}^\top \boldsymbol{L}^{(B)}\boldsymbol{U}) + \lambda_C \mathrm{trace}(\boldsymbol{V}^\top \boldsymbol{L}^{(C)}\boldsymbol{V}) \\ &+ \lambda(\|\boldsymbol{U}\|^2 + \|\boldsymbol{V}\|^2), \end{aligned} \tag{8.107}$$

where λ is a hyperparameter. The partial derivative of the Lagrangian function with respect to $\boldsymbol{U}$ ($\boldsymbol{V}$) can be the same as that of the input of two matrices, including a similarity matrix, when we ignore the terms, which are not related to the matrix parameter, which we focus on (for example when focusing on $\boldsymbol{U}$, fixing $\boldsymbol{V}$).

Then using (8.77), we can have the following multiplicative update rules for $\boldsymbol{U}$ and $\boldsymbol{V}$:

$$\boldsymbol{U}_{ik} \leftarrow \boldsymbol{U}_{ik}\frac{(\boldsymbol{A}\boldsymbol{V}+\lambda_B\boldsymbol{B}\boldsymbol{U})_{ik}}{(\boldsymbol{U}\boldsymbol{V}^\top\boldsymbol{V}+\lambda_B\boldsymbol{D}^{(B)}\boldsymbol{U}+\lambda\boldsymbol{U})_{ik}}, \tag{8.108}$$

$$\boldsymbol{V}_{jk} \leftarrow \boldsymbol{V}_{jk}\frac{(\boldsymbol{A}^\top\boldsymbol{U}+\lambda_C\boldsymbol{C}^\top\boldsymbol{V})_{jk}}{(\boldsymbol{V}\boldsymbol{U}^\top\boldsymbol{U}+\lambda_C\boldsymbol{D}^{(C)}\boldsymbol{V}+\lambda\boldsymbol{V})_{jk}}. \tag{8.109}$$

Finally **Procedure 8.12** shows a pseudocode of the above procedure for three matrices including two similarity matrices for both the instance and feature sides.

Procedure 8.12: COLLABORATIVE MATRIX FACTORIZATION FOR THREE MATRICES SHARING INSTANCES AND FEATURES, WHERE TWO MATRICES ARE SIMILARITY MATRICES.

Input: Three matrices: $\boldsymbol{A}$, $\boldsymbol{B}$ and $\boldsymbol{C}$; low-rank: K
Output: Low-rank matrices: $\boldsymbol{U}$ and $\boldsymbol{V}$

```
Initialization: Initialize U and V.;
repeat
    Update U by (8.109).;
    Update V by (8.109) .;
until convergence;
Output estimated Û and V̂.;
```

R Matrices, Sharing Instances with S Matrices and Features with T Matrices

We now think matrix factorization over a general input, which is shown in Fig. 8.13, as a schematic picture. This input has three types of matrices, where one type of matrices ($\boldsymbol{A}^{(1)}, \ldots, \boldsymbol{A}^{(R)}$) shares instances with another type of matrices ($\boldsymbol{B}^{(1)}, \ldots, \boldsymbol{B}^{(S)}$) and features with the other type of matrices ($\boldsymbol{C}^{(1)}, \ldots, \boldsymbol{C}^{(T)}$).

There are a lot of problem settings for this input, depending on the assumption behind the data. We think about the following three settings among them:

1. Only one set of low-rank matrices.

In this general setting, an assumption is the sizes of all $\boldsymbol{A}^{(*)}$ are all kept the same. Thus when we think $\boldsymbol{A}^{(*)}$ as a user-item matrix, this assumption means that we use the same set of users and the same set of items for all $\boldsymbol{A}^{(*)}$. This implies $\boldsymbol{A}^{(*)}$ can share the same factors or low-rank matrices, and then we here assume that all $\boldsymbol{A}^{(*)}$ are generated by the common pair of two low-rank matrices, i.e. $(\boldsymbol{U}, \boldsymbol{V})$. This idea can be described as below:

$$\boldsymbol{A}^{(r)} \sim \boldsymbol{U}\boldsymbol{V}^{\top} \ (r = 1, \ldots, R) \tag{8.110}$$

Then we can place the same assumption for $\boldsymbol{B}^{(*)}$ and $\boldsymbol{C}^{(*)}$. That is,

$$\boldsymbol{B}^{(s)} \sim \boldsymbol{U}\boldsymbol{P}^{\top} \ (s = 1, \ldots, S), \quad \boldsymbol{C}^{(t)} \sim \boldsymbol{Q}\boldsymbol{V}^{\top} \ (t = 1, \ldots, T), \tag{8.111}$$

Thus we use only four low-rank matrices for given multiple matrices, meaning that we just attempt to capture shared information of the input three types of matrices by only four low-rank matrices. Fig. 8.16 shows a schematic picture of only four low-rank matrices obtained by factorizing given three types of multiple matrices.

Following (8.96), this idea leads to the following formulation of the problem

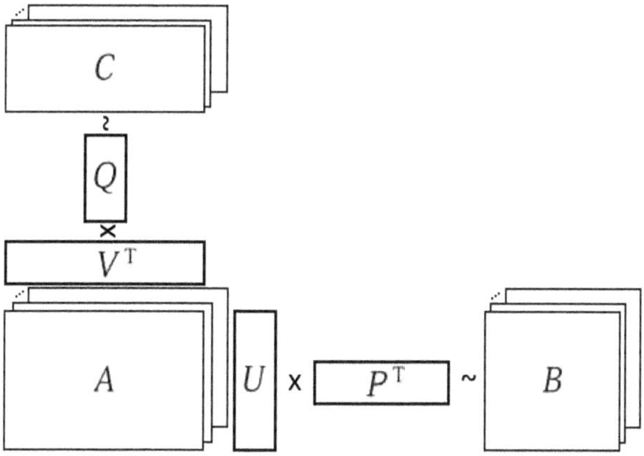

Figure 8.16: Collaborative matrix factorization of factorizing three types of matrices, sharing instances and features, into four low-rank matrices.

of estimating four low-rank matrices shared by the input multiple matrices:

$$\begin{aligned}\min_{\boldsymbol{U},\boldsymbol{V},\boldsymbol{P},\boldsymbol{Q}} \quad & \sum_r ||\boldsymbol{A}^{(r)} - \boldsymbol{U}\boldsymbol{V}^\top||^2 + \lambda_B \sum_s ||\boldsymbol{B}^{(s)} - \boldsymbol{U}\boldsymbol{P}^\top||^2 \\ + \quad & \lambda_C \sum_t ||\boldsymbol{C}^{(t)} - \boldsymbol{Q}\boldsymbol{V}^\top||^2 \\ \text{subject to} \quad & ||\boldsymbol{U}||^2 = \text{Const.}, \; ||\boldsymbol{V}||^2 = \text{Const.}, \\ & ||\boldsymbol{P}||^2 = \text{Const.}, \; ||\boldsymbol{Q}||^2 = \text{Const.}, \end{aligned} \tag{8.112}$$

where λ_B and λ_C are hyperparameters.

In fact we can notice that this leads to multiplicative update rules, which are very similar to those with the input of three matrices $\boldsymbol{A}$, $\boldsymbol{B}$ and $\boldsymbol{C}$ only. That is, we can obtain the multiplicative update rules just simply by replacing $\boldsymbol{A}$, $\boldsymbol{B}$, $\boldsymbol{C}$ with $\sum_r \boldsymbol{A}^{(r)}$, $\sum_s \boldsymbol{B}^{(s)}$, $\sum_t \boldsymbol{C}^{(t)}$, respectively, in (8.101) to (8.104), as follows:

$$\boldsymbol{U}_{ik} \leftarrow \boldsymbol{U}_{ik} \frac{((\sum_r \boldsymbol{A}^{(r)})\boldsymbol{V} + \lambda_B(\sum_s \boldsymbol{B}^{(s)})\boldsymbol{P})_{ik}}{(\boldsymbol{U}(\boldsymbol{V}^\top\boldsymbol{V} + \lambda_B\boldsymbol{P}^\top\boldsymbol{P} + \lambda\boldsymbol{I}))_{ik}}, \tag{8.113}$$

$$\boldsymbol{V}_{kj} \leftarrow \boldsymbol{V}_{kj} \frac{((\sum_r \boldsymbol{A}^{(r)})^\top\boldsymbol{U} + \lambda_C(\sum_r \boldsymbol{C}^{(t)})^\top\boldsymbol{Q})_{ik}}{(\boldsymbol{V}(\boldsymbol{U}^\top\boldsymbol{U} + \lambda_C\boldsymbol{Q}^\top\boldsymbol{Q} + \lambda\boldsymbol{I}))_{kj}}, \tag{8.114}$$

$$\boldsymbol{P}_{jk} \leftarrow \boldsymbol{P}_{jk} \frac{((\sum_s \boldsymbol{B}^{(s)})^\top\boldsymbol{U})_{jk}}{(\boldsymbol{P}(\boldsymbol{U}^\top\boldsymbol{U} + \lambda_P\boldsymbol{I}))_{jk}}, \tag{8.115}$$

$$\boldsymbol{Q}_{ik} \leftarrow \boldsymbol{Q}_{ik} \frac{((\sum_t \boldsymbol{C}^{(t)})\boldsymbol{V})_{ik}}{(\boldsymbol{Q}(\boldsymbol{V}^\top\boldsymbol{V} + \lambda_Q\boldsymbol{I}))_{ik}}, \tag{8.116}$$

where $\lambda_P = \frac{\lambda}{\lambda_B}$ and $\lambda_Q = \frac{\lambda}{\lambda_C}$.

This would be a very simple modification of the update rules for the general multiple matrix input, while this result would be very reasonable, because the

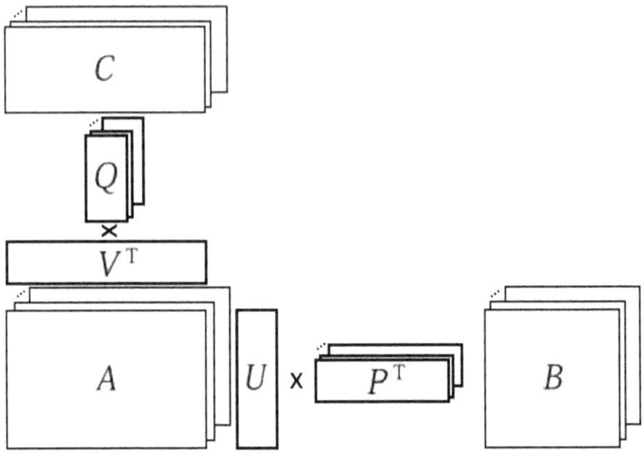

Figure 8.17: Collaborative matrix factorization of factorizing three types of multiple matrices, sharing instances and features, into partially multiple low-rank matrices.

obtained low-rank matrices should become a type of mean values of given inputs. However, we may say that this assumption over $\boldsymbol{U}, \boldsymbol{V}, \boldsymbol{P}$ and $\boldsymbol{Q}$ might be too simple for making the most of information of all given matrices.

2. Multiple low-rank matrices for $\boldsymbol{B}^{(*)}$ and $\boldsymbol{C}^{(*)}$.

A possible and reasonable direction in modification of the low-rank matrices of the above assumption is to increase the number of each low-rank matrix. The first step is to increase the low-rank matrices for $\boldsymbol{B}^{(*)}$ and $\boldsymbol{C}^{(*)}$, i.e. $\boldsymbol{P}^{(*)}$ and $\boldsymbol{Q}^{(*)}$ from $\boldsymbol{P}$ and $\boldsymbol{Q}$, respectively, and then the low-rank matrices for $\boldsymbol{A}^{(*)}$ are still kept the same: $\boldsymbol{U}$ and $\boldsymbol{V}$. That is,

$$\boldsymbol{A}^{(r)} \sim \boldsymbol{U}\boldsymbol{V}^{\top} \ (r = 1, \ldots, R) \tag{8.117}$$
$$\boldsymbol{B}^{(s)} \sim \boldsymbol{U}(\boldsymbol{P}^{(s)})^{\top} \ (s = 1, \ldots, S), \tag{8.118}$$
$$\boldsymbol{C}^{(t)} \sim \boldsymbol{Q}^{(t)}\boldsymbol{V}^{\top} \ (t = 1, \ldots, T). \tag{8.119}$$

Fig. 8.17 shows a schematic picture of this idea with $\boldsymbol{U}$, $\boldsymbol{V}$, $\boldsymbol{P}^{(*)}$ and $\boldsymbol{Q}^{(*)}$ to be obtained by factorizing given three types of multiple matrices.

The intuitive explanation behind this idea is that given information can be divided into two types: one is shared by given matrices, and the other is specific to each given matrix, and then the first information should be captured by $\boldsymbol{U}$ and $\boldsymbol{V}$ and the second information should be by $\boldsymbol{P}^{(*)}$ and $\boldsymbol{Q}^{(*)}$ for $\boldsymbol{B}^{(*)}$ and $\boldsymbol{C}^{(*)}$, respectively. Thus we can say that our matrix factorization divide given information into two types, and using the matrices which capture the two types of information separately, we can reproduce the original given input matrices.

We then formulate this problem as follows:

$$\begin{aligned}
\min_{U,V,P^{(*)},Q^{(*)}} \quad & \sum_r ||A^{(r)} - UV^\top||^2 \\
+ \quad & \lambda_B \sum_s ||B^{(s)} - U(P^{(s)})^\top||^2 \\
+ \quad & \lambda_C \sum_t ||C^{(t)} - Q^{(t)}V^\top||^2 \\
\text{subject to} \quad & ||U||^2 = \text{Const.},\ ||V||^2 = \text{Const.}, \\
& ||P^{(s)}||^2 = \text{Const.}\ (s = 1, \dots, S), \\
& ||Q^{(t)}||^2 = \text{Const.}\ (t = 1, \dots, T),
\end{aligned} \tag{8.120}$$

where λ_B and λ_C are hyperparameters.

In fact this formulation also leads to rather simple multiplicative update rules, which are slightly modified from the input of only single A, B and C and also from the single set of low-rank matrices, as follows:

$$U_{ik} \leftarrow U_{ik} \frac{((\sum_r A^{(r)})V + \lambda_B(\sum_s B^{(s)}P^{(s)}))_{ik}}{(U(V^\top V + \lambda_B(\sum_s (P^{(s)})^\top P^{(s)}) + \lambda I))_{ik}}, \tag{8.121}$$

$$V_{kj} \leftarrow V_{kj} \frac{((\sum_r A^{(r)})^\top U + \lambda_C(\sum_t (C^{(t)})^\top Q^{(t)}))_{ik}}{(V(U^\top U + \lambda_C(\sum_t (Q^{(t)})^\top Q^{(t)}) + \lambda I))_{kj}}, \tag{8.122}$$

$$P^{(s)}_{jk} \leftarrow P^{(s)}_{jk} \frac{((B^{(s)})^\top U)_{jk}}{(P^{(s)}(U^\top U + \lambda_P I))_{jk}} \quad (s = 1, \dots, S), \tag{8.123}$$

$$Q^{(t)}_{ik} \leftarrow Q^{(t)}_{ik} (\frac{C^{(t)}V)_{ik}}{(Q^{(t)}(V^\top V + \lambda_Q I))_{ik}} \quad (t = 1, \dots, T), \tag{8.124}$$

where $\lambda_P = \frac{\lambda}{\lambda_B}$ and $\lambda_Q = \frac{\lambda}{\lambda_C}$.

Procedure 8.13 shows a pseudocode of estimating parameters of the above setting.

3. Multiple low-rank matrices for all inputs.

Then the final assumption we can have is rather straightforward but somewhat reasonable as an extension from the above assumptions. That is, we further assume multiple low-rank matrices for $A^{(*)}$, i.e. $U^{(*)}$ and $V^{(*)}$. We can write them as follows:

$$A^{(r)} \sim U^{(r)}(V^{(r)})^\top \ (r = 1, \dots, R) \tag{8.125}$$

$$B^{(s)} \sim U^{(r)}(P^{(s)})^\top \ (r = 1, \dots, R, s = 1, \dots, S), \tag{8.126}$$

$$C^{(t)} \sim Q^{(t)}(V^{(r)})^\top \ (r = 1, \dots, R, t = 1, \dots, T). \tag{8.127}$$

Fig. 8.18 shows a schematic picture of this idea with all multiple low-rank matrices to be obtained by factorizing given three types of multiple matrices.

This idea indicates that each of all low-rank matrices is rather specific to the corresponding given matrix, while these given input-specific low-rank matrices are

Procedure 8.13: COLLABORATIVE MATRIX FACTORIZATION FOR THREE TYPES OF MATRICES SHARING INSTANCES AND FEATURES, WITH PARTIALLY MULTIPLE LOW-RANK MATRICES.

Input: Three types of matrices: $\boldsymbol{A}^{(r)}(r = 1, \ldots, R)$, $\boldsymbol{B}^{(s)}$ $(s = 1, \ldots, S)$ and $\boldsymbol{C}^{(t)}$ $(t = 1, \ldots, T)$; low-rank: K

Output: Low-rank matrices: $\boldsymbol{U}$, $\boldsymbol{V}$, $\boldsymbol{P}^{(s)}(s = 1, \ldots, S)$ and $\boldsymbol{Q}^{(t)}(t = 1, \ldots, T)$

```
Initialization: Initialize U, V, P^(s)(s = 1,...,S) and Q^(t)(t = 1,...,T).;
repeat
    Update U by (8.121).;
    Update V by (8.122).;
    for s ← 1 to S do
        Update P^(s) by (8.123).;
    for t ← 1 to T do
        Update Q^(t) by (8.124).;
until convergence;
Output estimated Û, V̂, P̂^(s)(s = 1,...,S) and Q̂^(t)(t = 1,...,T).;
```

balanced by the counterpart low-rank matrices, which are specific to the counterpart given matrix, so that the factorized low-rank matrices can reproduce all given matrices. For example, $\boldsymbol{U}^{(r)}$ should be specific to $\boldsymbol{A}^{(r)}$, while this specificity should be balanced by $\boldsymbol{B}^{(s)}$, since $\boldsymbol{U}^{(r)}$ is used to reproduce $\boldsymbol{B}^{s}$. This relationship can be applied to $\boldsymbol{V}^{(r)}$, $\boldsymbol{A}^{(r)}$ and $\boldsymbol{C}^{t}$ as well.

We can formulate this idea as follows:

$$\begin{aligned}
\min_{\boldsymbol{U}^{(*)},\boldsymbol{V}^{(*)},\boldsymbol{P}^{(*)},\boldsymbol{Q}^{(*)}} \quad & \sum_r ||\boldsymbol{A}^{(r)} - \boldsymbol{U}^{(r)}(\boldsymbol{V}^{(r)})^\mathsf{T}||^2 \\
+ \quad & \lambda_B \sum_r \sum_s ||\boldsymbol{B}^{(s)} - \boldsymbol{U}^{(r)}(\boldsymbol{P}^{(s)})^\mathsf{T}||^2 \\
+ \quad & \lambda_C \sum_r \sum_t ||\boldsymbol{C}^{(t)} - \boldsymbol{Q}^{(t)}(\boldsymbol{V}^{(r)})^\mathsf{T}||^2 \\
\text{subject to} \quad & ||\boldsymbol{U}^{(r)}||^2 = \text{Const.}\ (r = 1, \ldots, R), \\
& ||\boldsymbol{V}^{(r)}||^2 = \text{Const.}\ (r = 1, \ldots, R), \\
& ||\boldsymbol{P}^{(s)}||^2 = \text{Const.}\ (s = 1, \ldots, S), \\
& ||\boldsymbol{Q}^{(t)}||^2 = \text{Const.}\ (t = 1, \ldots, T), \qquad (8.128)
\end{aligned}$$

where λ_B and λ_C are hyperparameters.

This formulation leads to the following multiplicative update rules, which might be also similar to the above two settings:

$$\boldsymbol{U}_{ik}^{(r)} \leftarrow \boldsymbol{U}_{ik}^{(r)} \frac{(\boldsymbol{A}^{(r)}\boldsymbol{V}^{(r)} + \lambda_B(\sum_s \boldsymbol{B}^{(s)}\boldsymbol{P}^{(s)}))_{ik}}{(\boldsymbol{U}^{(r)}((\boldsymbol{V}^{(r)})^\mathsf{T}\boldsymbol{V}^{(r)} + \lambda_B(\sum_s (\boldsymbol{P}^{(s)})^\mathsf{T}\boldsymbol{P}^{(s)}) + \lambda \boldsymbol{I}))_{ik}} \quad (r = 1, \ldots, R), \qquad (8.129)$$

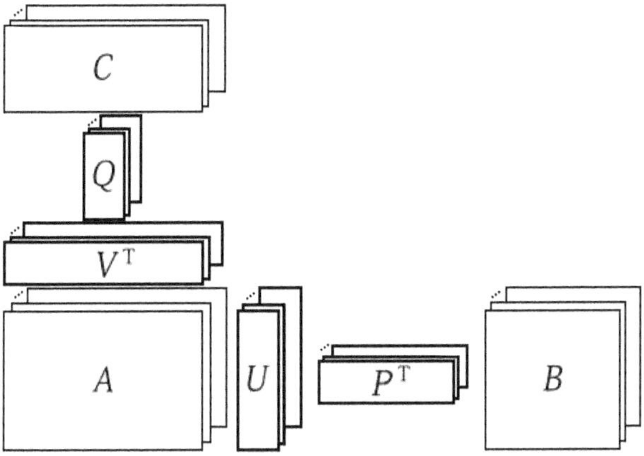

Figure 8.18: Collaborative matrix factorization of factorizing three types of multiple matrices, sharing instances and features, into all multiple low-rank matrices.

$$V_{kj}^{(r)} \leftarrow V_{kj}^{(r)} \frac{((\boldsymbol{A}^{(r)})^{\mathsf{T}}\boldsymbol{U}^{(r)} + \lambda_C(\sum_t(\boldsymbol{C}^{(t)})^{\mathsf{T}}\boldsymbol{Q}^{(t)}))_{ik}}{(\boldsymbol{V}^{(r)}((\boldsymbol{U}^{(r)})^{\mathsf{T}}\boldsymbol{U}^{(r)} + \lambda_C(\sum_t(\boldsymbol{Q}^{(t)})^{\mathsf{T}}\boldsymbol{Q}^{(t)}) + \lambda \boldsymbol{I}))_{kj}} \quad (r = 1, \ldots, R), \tag{8.130}$$

$$P_{jk}^{(s)} \leftarrow P_{jk}^{(s)} \frac{(\sum_r(\boldsymbol{B}^{(s)})^{\mathsf{T}}\boldsymbol{U}^{(r)})_{jk}}{(\boldsymbol{P}^{(s)}((\sum_r(\boldsymbol{U}^{(r)})^{\mathsf{T}}\boldsymbol{U}^{(r)}) + \lambda_P \boldsymbol{I}))_{jk}} \quad (s = 1, \ldots, S), \tag{8.131}$$

$$Q_{ik}^{(t)} \leftarrow Q_{ik}^{(t)} \frac{(\sum_r \boldsymbol{C}^{(t)}\boldsymbol{V}^{(r)})_{ik}}{(\boldsymbol{Q}^{(t)}((\sum_r(\boldsymbol{V}^{(r)})^{\mathsf{T}}\boldsymbol{V}^{(r)}) + \lambda_Q \boldsymbol{I}))_{ik}} \quad (t = 1, \ldots, T), \tag{8.132}$$

where $\lambda_P = \frac{\lambda}{\lambda_B}$ and $\lambda_Q = \frac{\lambda}{\lambda_C}$.

Procedure 8.14 shows a pseudocode of estimating parameters of the above setting.

A clear problem of this approach is we have a larger number of low-rank matrices than that of given input matrices. That is, the number of low-rank matrices reaches $2R + S + T$, which is larger than the number of input matrices, i.e. $R + S + T$. Although the rank of low-rank matrices can be forced to be much lower than those of the input matrices, this setting may have too many parameters, which may cause problems in optimization.

R Matrices, Sharing Instances with S Matrices and Features with T Matrices, where the S and T Matrices are Similarity Matrices

For the input of three matrices, $\boldsymbol{A}$, $\boldsymbol{B}$ and $\boldsymbol{C}$ only, when we have the assumption that $\boldsymbol{B}$ and $\boldsymbol{C}$ are similarity matrices, we could have a simpler solution. Similarly, for $\boldsymbol{A}^{(*)}$, $\boldsymbol{B}^{(*)}$ and $\boldsymbol{C}^{(*)}$, if $\boldsymbol{B}^{(*)}$ and $\boldsymbol{C}^{(*)}$ are similarity matrices, we can derive a simpler algorithm of estimating probability matrices. Under this similarity matrix

Procedure 8.14: COLLABORATIVE MATRIX FACTORIZATION FOR THREE TYPES OF MATRICES SHARING INSTANCES AND FEATURES, WHERE MULTIPLE LOW-RANK MATRICES ARE FOR ALL THREE TYPES OF MATRICES.

Input: Three types of matrices: $\boldsymbol{A}^{(r)} (r = 1, \ldots, R)$, $\boldsymbol{B}^{(s)}$ $(s = 1, \ldots, S)$ and $\boldsymbol{C}^{(t)}$ $(t = 1, \ldots, T)$; low-rank: K

Output: Low-rank matrices: $\boldsymbol{U}^{(r)}, \boldsymbol{V}^{(r)} (r = 1, \ldots, R)$, $\boldsymbol{P}^{(s)} (s = 1, \ldots, S)$ and $\boldsymbol{Q}^{(t)} (t = 1, \ldots, T)$

```
Initialization: Initialize U^(r), V^(r) (r = 1,...,R), P^(s) (s = 1,...,S) and
  Q^(t) (t = 1,...,T).;
repeat
    for r ← 1 to R do
        Update U^(r) by (8.129).;
        Update V^(r) by (8.130).;
    for s ← 1 to S do
        Update P^(s) by (8.131).;
    for t ← 1 to T do
        Update Q^(t) by (8.132).;
until convergence;
Output estimated Û^(r), V̂^(r), P̂^(s) (s = 1,...,S) and Q̂^(t) (t = 1,...,T).;
```

assumption, if $\boldsymbol{A}^*$ are all $(N \times M)$ matrices, the sizes of $\boldsymbol{B}^{(*)}$ and $\boldsymbol{C}^{(*)}$ are all $(N \times N)$ and $(M \times M)$, respectively. Then we have only two sizes of low-rank matrices, i.e. $(N \times K)$ and $(M \times K)$ for the $\boldsymbol{B}^{(*)}$ and $\boldsymbol{C}^{(*)}$ sides, respectively, where K is the rank of the low-rank matrices[7].

As we have considered three problem settings for the general case of the input, and we again consider the same three problem settings here.

1. Only one set of low-rank matrices.

The general case (with no similarity matrices) assumes that we have four low-rank matrices, i.e. $\boldsymbol{U}$, $\boldsymbol{V}$, $\boldsymbol{P}$ and $\boldsymbol{Q}$. Then if $\boldsymbol{B}^{(*)}$ and $\boldsymbol{C}^{(*)}$ are similarity matrices, $\boldsymbol{P} = \boldsymbol{U}$ and $\boldsymbol{Q} = \boldsymbol{V}$, resulting in only two matrices, $\boldsymbol{U}$ and $\boldsymbol{V}$.

Fig. 8.19 shows a schematic picture of this setting, with only two low-rank matrices for three different types of input matrices. Thus,

$$\begin{aligned} \boldsymbol{A}^{(r)} &\sim \boldsymbol{U}\boldsymbol{V}^\top \ (r = 1, \ldots, R), \\ \boldsymbol{B}^{(s)} &\sim \boldsymbol{U}\boldsymbol{U}^\top \ (s = 1, \ldots, S), \\ \boldsymbol{C}^{(t)} &\sim \boldsymbol{V}\boldsymbol{V}^\top \ (t = 1, \ldots, T). \end{aligned} \tag{8.133}$$

We have only two low-rank matrices, and the problem can be formulated as fol-

[7]We assume K common for low-rank matrices, while K can be different between the $\boldsymbol{B}^{(*)}$ and $\boldsymbol{C}^{(*)}$ sides.

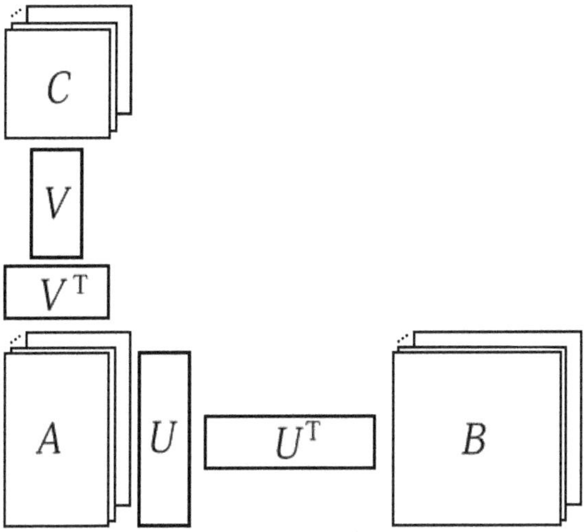

Figure 8.19: Collaborative matrix factorization of factorizing three types of multiple matrices, where two types of matrices are similarity matrices, into four low-rank matrices.

lows:

$$\begin{aligned}\min_{\boldsymbol{U},\boldsymbol{V}} \quad & \sum_r ||\boldsymbol{A}^{(r)} - \boldsymbol{U}\boldsymbol{V}^\top||^2 \\ & + \lambda_B \sum_s \mathrm{trace}(\boldsymbol{U}^\top \boldsymbol{L}^{(B,s)}\boldsymbol{U}) + \lambda_C \sum_t \mathrm{trace}(\boldsymbol{V}^\top \boldsymbol{L}^{(C,t)}\boldsymbol{V}) \\ \text{subject to} \quad & ||\boldsymbol{U}||^2 = \text{Const.}, \ ||\boldsymbol{V}||^2 = \text{Const.}, \end{aligned} \tag{8.134}$$

where $\boldsymbol{L}^{(B,s)}(= \boldsymbol{D}^{(B,s)} - \boldsymbol{B}^{(s)})$ and $\boldsymbol{L}^{(C,t)}(= \boldsymbol{D}^{(C,t)} - \boldsymbol{C}^{(t)})$ are graph Laplacian of $\boldsymbol{B}^{(s)}$ and $\boldsymbol{C}^{(t)}$, respectively, and $\boldsymbol{D}^{(B,s)}$ and $\boldsymbol{D}^{(C,t)}$ are diagonal matrices for $\boldsymbol{B}^{(s)}$ and $\boldsymbol{D}^{(C,t)}$, respectively, where the k-th diagonal elements $d_k^{(B,s)}$ and $d_k^{(C,t)}$ are computed as $d_k^{(B,s)} = \sum_i \boldsymbol{B}_{ki}^{(s)}$, $d_k^{(C,t)} = \sum_i \boldsymbol{C}_{ki}^{(t)}$, respectively, and λ_B and λ_C are hyperparameters.

Then following the derivation of (8.108) and (8.109), we can derive the following multiplicative update rules, which are a straightforward extension from (8.108) and (8.109).

$$\boldsymbol{U}_{ik} \leftarrow \boldsymbol{U}_{ik} \frac{((\sum_r \boldsymbol{A}^{(r)}\boldsymbol{V}) + \lambda_B(\sum_s \boldsymbol{B}^{(s)}\boldsymbol{U}))_{ik}}{(\boldsymbol{U}\boldsymbol{V}^\top\boldsymbol{V} + \lambda_B(\sum_s \boldsymbol{D}^{(B,s)}\boldsymbol{U}) + \lambda\boldsymbol{U})_{ik}}, \tag{8.135}$$

$$\boldsymbol{V}_{jk} \leftarrow \boldsymbol{V}_{jk} \frac{((\sum_r (\boldsymbol{A}^{(r)})^\top\boldsymbol{U}) + \lambda_C(\sum_t (\boldsymbol{C}^{(t)})^\top\boldsymbol{V}))_{jk}}{(\boldsymbol{V}\boldsymbol{U}^\top\boldsymbol{U} + \lambda_C(\sum_t \boldsymbol{D}^{(C,t)}\boldsymbol{V}) + \lambda\boldsymbol{V})_{jk}}, \tag{8.136}$$

They are very similar to (8.108) and (8.109). By this assumption, for example, all $\boldsymbol{A}^{(*)}$ are estimated as the same matrix by using $\boldsymbol{U}$ and $\boldsymbol{V}$, and so this model would be too simple, and we will move to the next problem setting.

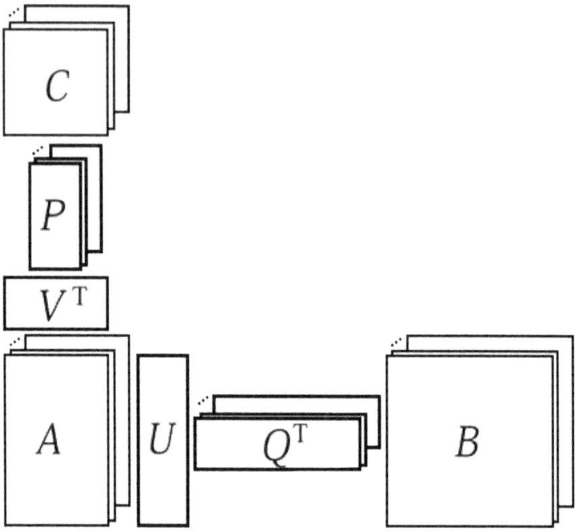

Figure 8.20: Collaborative matrix factorization of factorizing three types of multiple matrices, where two types of matrices are similarity matrices, into two low-rank matrices and two types of multiple low-rank matrices.

2. Multiple low-rank matrices for $\boldsymbol{B}^{(*)}$ and $\boldsymbol{C}^{(*)}$.

Then the next setting is to consider multiple low-rank matrices for the both $\boldsymbol{B}^{(*)}$ and $\boldsymbol{C}^{(*)}$ sides. That is, we have $\boldsymbol{P}^{(*)}$ and $\boldsymbol{Q}^{(*)}$. Then $\boldsymbol{P}^{(*)}$ cannot be necessarily the same each other and this is true of $\boldsymbol{Q}^{(*)}$, while the size of $\boldsymbol{P}^{(*)}$ and $\boldsymbol{Q}^{(*)}$ are the same as those of $\boldsymbol{U}$ and $\boldsymbol{V}$, respectively.

This approximation can be written as follows:

$$\begin{aligned} \boldsymbol{A}^{(r)} &\sim \boldsymbol{U}\boldsymbol{V}^\top \ (r = 1, \ldots, R), \\ \boldsymbol{B}^{(s)} &\sim \boldsymbol{U}(\boldsymbol{P}^{(s)})^\top \ (s = 1, \ldots, S), \\ \boldsymbol{C}^{(t)} &\sim \boldsymbol{Q}^{(t)}\boldsymbol{V}^\top \ (t = 1, \ldots, T). \end{aligned} \tag{8.137}$$

Also Fig. 8.20 shows a schematic picture of this matrix factorization for given three types of matrices with two types of similarity matrices into four types of low-rank matrices, in which two matrices of the $\boldsymbol{B}^{(*)}$ and $\boldsymbol{C}^{(*)}$ sides are multiple matrices.

We can then write the objective function by using graph Laplacian for the

$\boldsymbol{B}^{(*)}$ and $\boldsymbol{C}^{(*)}$ sides:

$$\begin{aligned}
\min_{\boldsymbol{U},\boldsymbol{V},\boldsymbol{P}^{(*)},\boldsymbol{Q}^{(*)}} \quad & \sum_r ||\boldsymbol{A}^{(r)} - \boldsymbol{U}\boldsymbol{V}^\top||^2 \\
& + \lambda_B \sum_s \mathrm{trace}(\boldsymbol{U}^\top \boldsymbol{L}^{(B,s)} \boldsymbol{P}^{(s)}) \\
& + \lambda_C \sum_t \mathrm{trace}(\boldsymbol{V}^\top \boldsymbol{L}^{(C,t)} \boldsymbol{Q}^{(t)}) \\
\text{subject to} \quad & ||\boldsymbol{U}||^2 = \text{Const.},\ ||\boldsymbol{V}||^2 = \text{Const.}, \\
& ||\boldsymbol{P}^{(s)}||^2 = \text{Const.}\ (s = 1, \dots, S), \\
& ||\boldsymbol{Q}^{(t)}||^2 = \text{Const.}\ (t = 1, \dots, T), \qquad (8.138)
\end{aligned}$$

where graph Laplacian $\boldsymbol{L}^{(B,s)}$ and $\boldsymbol{L}^{(C,t)}$ are the same as those in the previous setting. Also note that the second and third terms have the following properties, because of the matrix trace and also low-rank matrices have the same size between $\boldsymbol{U}$ ($\boldsymbol{V}$) and $\boldsymbol{P}^{(*)}$ ($\boldsymbol{Q}^{(*)}$):

$$\mathrm{trace}(\boldsymbol{U}^\top \boldsymbol{L}^{(B,s)} \boldsymbol{P}^{(s)}) = \mathrm{trace}((\boldsymbol{P}^{(s)})^\top \boldsymbol{L}^{(B,s)} \boldsymbol{U}), \qquad (8.139)$$
$$\mathrm{trace}(\boldsymbol{V}^\top \boldsymbol{L}^{(C,t)} \boldsymbol{Q}^{(t)}) = \mathrm{trace}((\boldsymbol{Q}^{(t)})^\top \boldsymbol{L}^{(C,t)} \boldsymbol{V}). \qquad (8.140)$$

Then first, update rules for $\boldsymbol{U}$ and $\boldsymbol{V}$ can be an extension from (8.135) and (8.136), as follows:

$$\boldsymbol{U}_{ik} \leftarrow \boldsymbol{U}_{ik} \frac{((\sum_r \boldsymbol{A}^{(r)}\boldsymbol{V}) + \lambda_B(\sum_s \boldsymbol{B}^{(s)}\boldsymbol{P}^{(s)}))_{ik}}{(\boldsymbol{U}\boldsymbol{V}^\top\boldsymbol{V} + \lambda_B(\sum_s \boldsymbol{D}^{(B,s)}\boldsymbol{P}^{(s)}) + \lambda\boldsymbol{U})_{ik}}, \qquad (8.141)$$
$$\boldsymbol{V}_{jk} \leftarrow \boldsymbol{V}_{jk} \frac{((\sum_r (\boldsymbol{A}^{(r)})^\top\boldsymbol{U}) + \lambda_C(\sum_t (\boldsymbol{C}^{(t)})^\top\boldsymbol{Q}^{(t)}))_{jk}}{(\boldsymbol{V}\boldsymbol{U}^\top\boldsymbol{U} + \lambda_C(\sum_t \boldsymbol{D}^{(C,t)}\boldsymbol{Q}^{(t)}) + \lambda\boldsymbol{V})_{jk}}, \qquad (8.142)$$

Next $\boldsymbol{P}^{(*)}$ ($\boldsymbol{Q}^{(*)}$) appears only two terms in (8.138), and so the update rule of $\boldsymbol{P}^{(*)}$ and $\boldsymbol{Q}^{(*)}$ are very simple, as follows:

$$\boldsymbol{P}^{(s)} \leftarrow \lambda_P \boldsymbol{L}^{(B,s)}\boldsymbol{U} \quad (s = 1, \dots, S), \qquad (8.143)$$
$$\boldsymbol{Q}^{(t)} \leftarrow \lambda_Q \boldsymbol{L}^{(C,t)}\boldsymbol{V} \quad (t = 1, \dots, T), \qquad (8.144)$$

where λ_P and λ_Q are hyperparameters.

Procedure 8.15 is a pseudocode of the above iterative procedure. In fact this pseudocode is totally the same as **Procedure 8.13**, which is the general case, which does not necessarily have similarity matrices.

3. Multiple low-rank matrices for all inputs.

The final assumption is that the four types of low rank matrices are all specific to given inputs, as follows:

$$\boldsymbol{A}^{(r)} \sim \boldsymbol{U}^{(r)}(\boldsymbol{V}^{(r)})^\top \ (r = 1, \dots, R) \qquad (8.145)$$
$$\boldsymbol{B}^{(s)} \sim \boldsymbol{U}^{(r)}(\boldsymbol{P}^{(s)})^\top \ (r = 1, \dots, R, s = 1, \dots, S), \qquad (8.146)$$
$$\boldsymbol{C}^{(t)} \sim \boldsymbol{Q}^{(t)}(\boldsymbol{V}^{(r)})^\top \ (r = 1, \dots, R, t = 1, \dots, T). \qquad (8.147)$$

Procedure 8.15: COLLABORATIVE MATRIX FACTORIZATION FOR THREE TYPES OF MATRICES SHARING INSTANCES AND FEATURES, WHERE TWO MATRICES ARE SIMILARITY MATRICES, WITH PARTIALLY MULTIPLE LOW-RANK MATRICES.

Input: Three types of matrices: $\boldsymbol{A}^{(r)}(r = 1, \ldots, R)$, $\boldsymbol{B}^{(s)}$ $(s = 1, \ldots, S)$ and $\boldsymbol{C}^{(t)}$ $(t = 1, \ldots, T)$; low-rank: K

Output: Low-rank matrices: $\boldsymbol{U}$, $\boldsymbol{V}$, $\boldsymbol{P}^{(s)}(s = 1, \ldots, S)$ and $\boldsymbol{Q}^{(t)}(t = 1, \ldots, T)$

```
Initialization: Initialize U, V, P^(s)(s = 1,...,S) and Q^(t)(t = 1,...,T).;
repeat
    Update U by (8.141).;
    Update V by (8.142).;
    for s ← 1 to S do
        Update P^(s) by (8.143).;
    for t ← 1 to T do
        Update Q^(t) by (8.144).;
until convergence;
Output estimated Û, V̂, P̂^(s)(s = 1,...,S) and Q̂^(t)(t = 1,...,T).;
```

Again this assumption is the same as the general case. Fig. 8.21 shows a schematic picture of given three types of multiple matrices including two types of similarity matrices and four types of multiple low-rank matrices.

Then we can formulate this problem setting as the following objective function:

$$\begin{aligned}
\min_{\boldsymbol{U}^{(*)},\boldsymbol{V}^{(*)},\boldsymbol{P}^{(*)},\boldsymbol{Q}^{(*)}} \quad & \sum_{r} ||\boldsymbol{A}^{(r)} - \boldsymbol{U}^{(r)}(\boldsymbol{V}^{(r)})^{\mathsf{T}}||^2 \\
+ \quad & \lambda_B \sum_{r} \sum_{s} \mathrm{trace}((\boldsymbol{U}^{(r)})^{\mathsf{T}} \boldsymbol{L}^{(B,s)} \boldsymbol{P}^{(s)}) \\
+ \quad & \lambda_C \sum_{r} \sum_{t} \mathrm{trace}((\boldsymbol{V}^{(r)})^{\mathsf{T}} \boldsymbol{L}^{(C,t)} \boldsymbol{Q}^{(t)}) \\
\text{subject to} \quad & ||\boldsymbol{U}^{(r)}||^2 = \text{Const. } (r = 1, \ldots, R), \\
& ||\boldsymbol{V}^{(r)}||^2 = \text{Const. } (r = 1, \ldots, R), \\
& ||\boldsymbol{P}^{(s)}||^2 = \text{Const. } (s = 1, \ldots, S), \\
& ||\boldsymbol{Q}^{(t)}||^2 = \text{Const. } (t = 1, \ldots, T),
\end{aligned} \tag{8.148}$$

where λ_B and λ_C are hyperparameters.

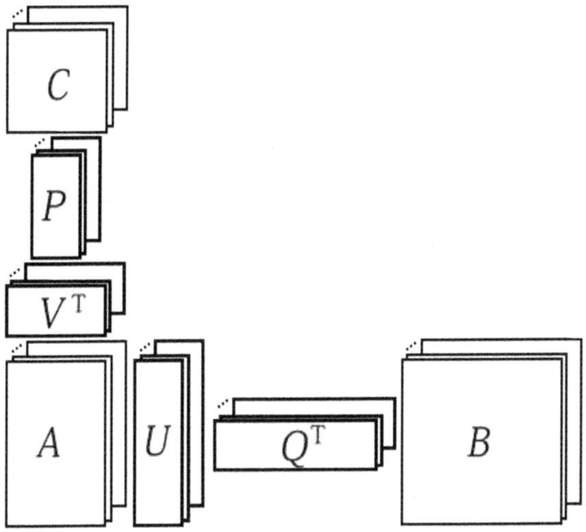

Figure 8.21: Collaborative matrix factorization of factorizing three types of multiple matrices, where two types of matrices are similarity matrices, into four types of multiple low-rank matrices.

Then the update rules can be given as follows:

$$U^{(r)}_{ik} \leftarrow U^{(r)}_{ik}\frac{((\boldsymbol{A}^{(r)}\boldsymbol{V}^{(r)}) + \lambda_B(\sum_s \boldsymbol{B}^{(s)}\boldsymbol{P}^{(s)}))_{ik}}{(\boldsymbol{U}^{(r)}(\boldsymbol{V}^{(r)})^{\mathsf{T}}\boldsymbol{V}^{(r)} + \lambda_B(\sum_s \boldsymbol{D}^{(B,s)}\boldsymbol{P}^{(s)}) + \lambda\boldsymbol{U}^{(r)})_{ik}} \quad (r = 1,\dots,R), \tag{8.149}$$

$$V^{(r)}_{jk} \leftarrow V^{(r)}_{jk}\frac{(((\boldsymbol{A}^{(r)})^{\mathsf{T}}\boldsymbol{U}^{(r)}) + \lambda_C(\sum_t (\boldsymbol{C}^{(t)})^{\mathsf{T}}\boldsymbol{Q}^{(t)}))_{jk}}{(\boldsymbol{V}^{(r)}(\boldsymbol{U}^{(r)})^{\mathsf{T}}\boldsymbol{U}^{(r)} + \lambda_C(\sum_t \boldsymbol{D}^{(C,t)}\boldsymbol{Q}^{(t)}) + \lambda\boldsymbol{V}^{(r)})_{jk}} \quad (r = 1,\dots,R), \tag{8.150}$$

$$\boldsymbol{P}^{(s)} \leftarrow \lambda_P \sum_r \boldsymbol{L}^{(B,s)}\boldsymbol{U}^{(r)} \quad (s = 1,\dots,S), \tag{8.151}$$

$$\boldsymbol{Q}^{(t)} \leftarrow \lambda_Q \sum_r \boldsymbol{L}^{(C,t)}\boldsymbol{V}^{(r)} \quad (t = 1,\dots,T), \tag{8.152}$$

where λ_P and λ_Q are hyperparameters.

Pseudocode 8.16 is a pseudocode of the procedure at this setting. The flow of this procedure is also totally the same as **Procedure 8.14**.

Again the problem of this setting is, as pointed out in the general case, we have a larger number of matrix parameters than the given input. That is the number of low-rank matrices reaches $2R+S+T$, which is larger than $R+S+T$, which is the number of input matrices. Of course the rank of low-rank matrices would be much smaller than the dimensions of the input matrices, and the total number of parameters would be smaller than the input instances, while this large number of parameters might cause some problem in optimization.

Procedure 8.16: COLLABORATIVE MATRIX FACTORIZATION OF FACTORIZING THREE TYPES OF MULTIPLE MATRICES, WHERE TWO TYPES OF MATRICES ARE SIMILARITY MATRICES, INTO FOUR TYPES OF MULTIPLE LOW-RANK MATRICES.

Input: Three types of matrices: $\boldsymbol{A}^{(r)}(r = 1, \dots, R)$, $\boldsymbol{B}^{(s)}$ $(s = 1, \dots, S)$ and $\boldsymbol{C}^{(t)}$ $(t = 1, \dots, T)$; low-rank: K

Output: Low-rank matrices: $\boldsymbol{U}^{(r)}, \boldsymbol{V}^{(}r)(r = 1, \dots, R)$, $\boldsymbol{P}^{(s)}(s = 1, \dots, S)$ and $\boldsymbol{Q}^{(t)}(t = 1, \dots, T)$

```
Initialization: Initialize U^(r), V^(r) (r = 1,...,R), P^(s) (s = 1,...,S) and
  Q^(t) (t = 1,...,T).;
repeat
    for r ← 1 to R do
        Update U^(r) by (8.149).;
        Update V^(r) by (8.150).;
    for s ← 1 to S do
        Update P^(s) by (8.151).;
    for t ← 1 to T do
        Update Q^(t) by (8.152).;
until convergence;
Output estimated Û^(r), V̂^(r), P̂^(s) (s = 1,...,S) and Q̂^(t) (t = 1,...,T).;
```

8.4 Notes

In this chapter, we have examined a variety of settings, in which we always have considered more than one matrices. In particular, our setting has multiple matrices as input, such as $R + S + T$ matrices in Section 8.3. Due to the large number of matrices as input, the number of parameters is also large, resulting in that the the problem of optimizing parameters for multiple matrices cannot be a simple convex problem. The solution we have shown for non-convex problems is an alternating optimization algorithm, based on gradient descent, which is entirely rather basic. As increasing the number of parameters, such as low-rank matrices, this type of algorithm would needs more time and also would not necessarily reach a good solution, since only a local minimum is guaranteed. Practically we need limit the size of parameters. In fact developing practically efficient and effective optimization algorithms over a larger number of parameters would be the most actively conducted research topic in the current machine learning research.

Bibliography

[1] (2013). https://www.ama.org/AboutAMA/Pages/Definition-of-Marketing.aspx.

[2] Amin, A., Al-Obeidat, F., Shah, B., Adnan, A., Loo, J., & Anwar, S. (2019). *Journal of Business Research,* **94**, 290 – 301.

[3] Ansari, A., Li, Y., & Zhang, J. Z. (2018). *Marketing Science,* **37** (6), 987–1008.

[4] Antons, D. & Breidbach, C. F. (2018). *Journal of Service Research,* **21** (1), 17–39.

[5] Armstrong, G. & Kotler, P. (2005). *Marketing: An Introduction.* Upper Saddle River, NJ: Pearson/Prentice Hall.

[6] Arora, A., Bansal, S., Kandpal, C., Aswani, R., & Dwivedi, Y. (2019). *Journal of Retailing and Consumer Services,* **49**, 86 – 101.

[7] Ascarza, E. (2018). *Journal of Marketing Research,* **55** (1), 80–98.

[8] Attenberg, J. & Ertekin, . (2013). *Class Imbalance and Active Learning* chapter 6, pp. 101–149. Oxford, UK: John Wiley & Sons, Ltd.

[9] Bag, S., Tiwari, M. K., & Chan, F. T. (2019). *Journal of Business Research,* **94**, 408 – 419.

[10] Bendixen, M. T. (1995). *Journal of Marketing Management,* **11** (6), 571–581.

[11] Berger, P. D., Eechambadi, N., George, M., Lehmann, D. R., Rizley, R., & Venkatesan, R. (2006). *Journal of Service Research,* **9** (2), 156–167.

[12] Berger, P. D. & Nasr, N. I. (1998). *Journal of Interactive Marketing,* **12** (1), 17 – 30.

[13] Bijmolt, T. H. A., Leeflang, P. S. H., Block, F., Eisenbeiss, M., Hardie, B. G. S., Lemmens, A., & Saffert, P. (2010). *Journal of Service Research,* **13** (3), 341–356.

[14] Blum, A. & Mitchell, T. (1998). In: *Proceedings of the Eleventh Annual Conference on Computational Learning Theory,* COLT' 98, pp. 92–100, New York, NY, USA: ACM.

[15] Borah, A. & Tellis, G. J. (2016). *Journal of Marketing Research,* **53** (2), 143–160.

[16] Bradlow, E. T., Gangwar, M., Kopalle, P., & Voleti, S. (2017). *Journal of Retailing,* **93** (1), 79 – 95. The Future of Retailing.

[17] Breiman, L. (1996). *Machine Learning,* **24** (2), 123–140.

[18] Cadez, I., Heckerman, D., Meek, C., Smyth, P., & White, S. (2003). *Data Mining and Knowledge Discovery,* **7** (4), 399–424.

[19] Carroll, J. D., Green, P. E., & Schaffer, C. M. (1986). *Journal of Marketing Research,* **23** (3), 271–280.

[20] Castéran, H., Meyer-Waarden, L., & Reinartz, W. (2017). *Modeling Customer Lifetime Value, Retention, and Churn* pp. 1–33. Cham: Springer International Publishing.

[21] Chawla, N. V., Bowyer, K. W., Hall, L. O., & Kegelmeyer, W. P. (2002). *J. Artif. Int. Res.*, **16** (1), 321–357.

[22] Chen, Y., Qi, Y., Liu, Q., & Chien, P. (2018). *Quantitative Marketing and Economics,* **16** (4), 409–440.

[23] Chintagunta, P., Hanssens, D. M., & Hauser, J. R. (2016). *Marketing Science,* **35** (3), 341–342.

[24] Costa, A., Guerreiro, J., Moro, S., & Henriques, R. (2019). *Journal of Retailing and Consumer Services,* **47**, 272 – 281.

[25] DeSarbo, W. S., Atalay, A. S., LeBaron, D., & Blanchard, S. J. (2008). *Journal of Consumer Research,* **35** (1), 142–153.

[26] Dhar, T. & Weinberg, C. B. (2016). *International Journal of Research in Marketing,* **33** (2), 392 – 408. The Entertainment Industry.

[27] Dolnicar, S. (2003). *Australasian Journal of Market Research,* **11** (2), 5–12.

[28] Dolnicar, S., Grün, B., & Leisch, F. (2018). *Market Segmentation Analysis: Understanding It, Doing It, and Making It Useful.* Management for Professionals. Singapore: Springer Singapore.

[29] Drucker, P. (1974). *Management: Tasks, Responsibilities, Practices.* San Francisco, CA: Harper & Row.

[30] Ellson, T. (2004). *Segmentation, Targeting, and Positioning* pp. 21–34. London: Palgrave Macmillan UK.

[31] Ertekin, S., Huang, J., & Giles, C. L. (2007). In: *Proceedings of the 30th Annual International ACM SIGIR Conference on Research and Development in Information Retrieval,* SIGIR '07, pp. 823–824, New York, NY, USA: ACM.

[32] Fan, Z.-P., Che, Y.-J., & Chen, Z.-Y. (2017). *Journal of Business Research,* **74**, 90 – 100.

[33] Farris, P., Bendle, N., Pfeifer, P., & Reibstein, D. (2010). *Marketing Metrics: The Definitive Guide to Measuring Marketing Performance.* London, England: Pearson Education.

[34] Felbermayr, A. & Nanopoulos, A. (2016). *Journal of Interactive Marketing,* **36**, 60 – 76.

[35] Glushko, R. J. & Nomorosa, K. J. (2013). *Journal of Service Research,* **16** (1), 21–38.

[36] Gönen, M. & Alpaydin, E. (2011). *J. Mach. Learn. Res.* **12**, 2211–2268.

[37] Gummesson, E. (1999). *Total Relationship Marketing.* Boston, MA: Butterworth-Heinemann.

[38] Gupta, S., Hanssens, D., Hardie, B., Kahn, W., Kumar, V., Lin, N., Ravishanker, N., & Sriram, S. (2006). *Journal of Service Research,* **9** (2), 139–155.

[39] Ha, K., Cho, S., & MacLachlan, D. (2005). *Journal of Interactive Marketing,* **19** (1), 17 – 30.

[40] Hartmann, J., Huppertz, J., Schamp, C., & Heitmann, M. (2019). *International Journal of Research in Marketing,* **36** (1), 20 – 38.

[41] Hauser, J. R. & Koppelman, F. S. (1979). *Journal of Marketing Research,* **16** (4), 495–506.

[42] He, H. & Ma, Y. (2013). *Imbalanced Learning: Foundations, Algorithms, and Applications.* Oxford, UK: Wiley-IEEE Press, 1st edition.

[43] Hoffman, D. L. & Franke, G. R. (1986). *Journal of Marketing Research,* **23** (3), 213–227.

[44] Homburg, C., Ehm, L., & Artz, M. (2015). *Journal of Marketing Research,* **52** (5), 629–641.

[45] Huang, D. & Luo, L. (2016). *Marketing Science,* **35** (3), 445–464.

[46] Huang, M.-H. & Rust, R. T. (2018). *Journal of Service Research,* **21** (2), 155–172.

[47] Huang, Z., Zeng, D., & Chen, H. (2007). *IEEE Intelligent Systems,* **22** (5), 68–78.

[48] Humphreys, A. & Wang, R. J.-H. (2017). *Journal of Consumer Research,* **44** (6), 1274–1306.

[49] Ilhan, B. E., Kbler, R. V., & Pauwels, K. H. (2018). *Journal of Interactive Marketing,* **43**, 33 – 51.

[50] Jahromi, A. T., Stakhovych, S., & Ewing, M. (2014). *Industrial Marketing Management,* **43** (7), 1258 – 1268.

[51] Jalali, N. Y. & Papatla, P. (2016). *Quantitative Marketing and Economics,* **14** (4), 353–384.

[52] Kamakura, W., Mela, C. F., Ansari, A., Bodapati, A., Fader, P., Iyengar, R., Naik, P., Neslin, S., Sun, B., Verhoef, P. C., Wedel, M., & Wilcox, R. (2005). *Marketing Letters,* **16** (3), 279–291.

[53] Kappe, E., Blank, A. S., & DeSarbo, W. S. (2018). *International Journal of Research in Marketing,* **35** (3), 415 – 431.

[54] Kloft, M., Brefeld, U., Sonnenburg, S., & Zien, A. (2011). *J. Mach. Learn. Res.* **12**, 953–997.

[55] Klostermann, J., Plumeyer, A., Bger, D., & Decker, R. (2018). *International Journal of Research in Marketing,* **35** (4), 538 – 556.

[56] Kotler, P. & Keller, K. (2011). *Marketing Management.* Upper Saddle River, NJ: Prentice Hall PTR.

[57] Kumar, V. (2018). *Journal of Marketing,* **82** (4), 1–12.

[58] Lauterborn, B. (1990). *Advertising Age,* **61** (41), 26.

[59] Learned, E. P., Christensen, C. R., Andrews, K., & Book, W. D. (1969). *Business Policy: Text and Cases.* Chicago, IL, USA: R. D. Irwin.

[60] Lemmens, A. & Croux, C. (2006). *Journal of Marketing Research,* **43** (2), 276–286.

[61] Lemon, K. N. & Verhoef, P. C. (2016). *Journal of Marketing,* **80** (6), 69–96.

[62] Levitt, T. (1960). *Harvard Business Review,* **38**, 45–56.

[63] Li, H. (2011). *IEICE Transactions on Information and Systems,* **94**, 1854–1862.

[64] Linoff, G. & Berry, M. (2011). *Data Mining Techniques: For Marketing, Sales, and Customer Relationship Management.* BusinessPro collection. Oxford, UK: Wiley.

[65] Liu, T.-Y. (2011). *Learning to Rank for Information Retrieval.* Berlin: Springer-Verlag Berlin Heidelberg.

[66] Liu, X., Burns, A. C., & Hou, Y. (2017). *Journal of Advertising,* **46** (2), 236–247.

[67] Liu, X., Singh, P. V., & Srinivasan, K. (2016). *Marketing Science,* **35** (3), 363–388.

[68] Liu-Thompkins, Y. & Malthouse, E. C. (2017). *Journal of Advertising,* **46** (1), 213–225.

[69] Malthouse, E. C. (1999). *Journal of Interactive Marketing,* **13** (4), 10 – 23.

[70] Mamitsuka, H. (2018). *Textbook of Machine Learning and Data Mining: with Bioinformatics Applications.* Kyoto, Japan: Global Data Science Publishing.

[71] Mamitsuka, H. & Abe, N. (2000). In: *Proceedings of the Seventeenth International Conference on Machine Learning* ICML '00 pp. 575–582, San Francisco, CA, USA: Morgan Kaufmann Publishers Inc.

[72] Mamitsuka, H., Okuno, Y., & Yamaguchi, A. (2003). *SIGKDD Explor. Newsl.* **5** (2), 113–121.

[73] McCarthy, J. E. (1964). *Basic Marketing: A Managerial Approach.* Homewood, IL: Irwin.

[74] Misra, K., Schwartz, E. M., & Abernethy, J. (2019). *Marketing Science,* **38** (2), 226–252.

[75] Moon, S. & Kamakura, W. A. (2017). *International Journal of Research in Marketing,* **34** (1), 265 – 285.

[76] Nam, H., Joshi, Y. V., & Kannan, P. (2017). *Journal of Marketing,* **81** (4), 88–108.

[77] Neslin, S. A., Gupta, S., Kamakura, W., Lu, J., & Mason, C. H. (2006). *Journal of Marketing Research,* **43** (2), 204–211.

[78] Ordenes, F. V., Theodoulidis, B., Burton, J., Gruber, T., & Zaki, M. (2014). *Journal of Service Research,* **17** (3), 278–295.

[79] Prahalad, C. K. & Hamel, G. (1990). *Harvard Business Review,* **68** (May-June), 79–91.

[80] Puranam, D., Narayan, V., & Kadiyali, V. (2017). *Marketing Science,* **36** (5), 726–746.

[81] Rabiner, L. R. (1989). *Proceedings of the IEEE,* **77** (2), 257–286.

[82] Ramaswamy, V. & Ozcan, K. (2018). *Journal of Marketing,* **82** (4), 19–31.

[83] Risselada, H., Verhoef, P. C., & Bijmolt, T. H. (2010). *Journal of Interactive Marketing,* **24** (3), 198 – 208.

[84] Romero, J., van der Lans, R., & Wierenga, B. (2013). *Journal of Interactive Marketing,* **27** (3), 185 – 208.

[85] Salminen, J., Yoganathan, V., Corporan, J., Jansen, B. J., & Jung, S.-G. (2019). *Journal of Business Research,* **101**, 203 – 217.

[86] Saunders, J. (1980). *European Journal of Marketing,* **14** (7), 422–435.

[87] Sawhney, M. (1999). *Business 2.0,* 116–121.

[88] Sawhney, M., Balasubramanian, S., & Krishnan, V. V. (2004). *MIT Sloan Management Review,* **45** (2), 34–43.

[89] Schapire, R. E. & Freund, Y. (2012). *Boosting: Foundations and Algorithms.* Cambridge, MA, USA: The MIT Press.

[90] Schwartz, E. M., Bradlow, E. T., & Fader, P. S. (2014). *Marketing Science,* **33** (2), 188–205.

[91] Schwartz, E. M., Bradlow, E. T., & Fader, P. S. (2017). *Marketing Science,* **36** (4), 500–522.

[92] Schweidel, D. A. & Moe, W. W. (2014). *Journal of Marketing Research,* **51** (4), 387–402.

[93] Seim, K. & Sinkinson, M. (2016). *Quantitative Marketing and Economics,* **14** (2), 129–155.

[94] Shawe-Taylor, J. & Cristianini, N. (2004). *Kernel Methods for Pattern Analysis.* New York, NY, USA: Cambridge University Press.

[95] Silva, A. T., Moro, S., Rita, P., & Cortez, P. (2018). *Journal of Retailing and Consumer Services,* **43**, 311 – 324.

[96] Singh, J. P., Irani, S., Rana, N. P., Dwivedi, Y. K., Saumya, S., & Roy, P. K. (2017). *Journal of Business Research,* **70**, 346 – 355.

[97] Syam, N. & Sharma, A. (2018). *Industrial Marketing Management,* **69**, 135 – 146.

[98] Tamaddoni, A., Stakhovych, S., & Ewing, M. (2016). *Journal of Service Research,* **19** (2), 123–141.

[99] Tehrani, A. F. & Ahrens, D. (2016). *Journal of Retailing and Consumer Services,* **32**, 131 – 138.

[100] Tillmanns, S., Hofstede, F. T., Krafft, M., & Goetz, O. (2017). *Journal of Marketing,* **81** (2), 99–113.

[101] Timoshenko, A. & Hauser, J. R. (2019). *Marketing Science,* **38** (1), 1–20.

[102] Tirunillai, S. & Tellis, G. J. (2014). *Journal of Marketing Research,* **51** (4), 463–479.

[103] Toubia, O., Iyengar, G., Bunnell, R., & Lemaire, A. (2019). *Journal of Marketing Research,* **56** (1), 18–36.

[104] Trusov, M., Ma, L., & Jamal, Z. (2016). *Marketing Science,* **35** (3), 405–426.

[105] Tuma, M. N., Decker, R., & Scholz, S. W. (2011). *International Journal of Market Research,* **53** (3), 391–414.

[106] Veropoulos, K., Campbell, C., & Cristianini, N. (1999). In: *Proceedings of the International Joint Conference on AI* pp. 55–60, San Francisco, CA, USA: Morgan Kaufmann Publishers.

[107] Verstraete, G., Aghezzaf, E.-H., & Desmet, B. (2019). *Journal of Retailing and Consumer Services,* **48**, 169 – 177.

[108] Villarroel Ordenes, F., Grewal, D., Ludwig, S., Ruyter, K. D., Mahr, D., & Wetzels, M. (2018). *Journal of Consumer Research,* **45** (5), 988–1012.

[109] Villarroel Ordenes, F., Ludwig, S., de Ruyter, K., Grewal, D., & Wetzels, M. (2017). *Journal of Consumer Research,* **43** (6), 875–894.

[110] Voorveld, H. A. (2019). *Journal of Advertising,* **48** (1), 14–26.

[111] Wedel, M. & Kannan, P. (2016). *Journal of Marketing,* **80** (6), 97–121.

[112] Yi, Z. (2018). In: *Marketing Services and Resources in Information Organizations,* (Yi, Z., ed) Chandos Information Professional Series pp. 39 – 48. Chandos Publishing Sawston, Cambridge, UK.

[113] Ying, Y., Feinberg, F., & Wedel, M. (2006). *Journal of Marketing Research,* **43** (3), 355–365.

[114] Zhang, Y. & Godes, D. (2018). *Marketing Science,* **37** (3), 425–444.

[115] Zheng, X., Ding, H., Mamitsuka, H., & Zhu, S. (2013). In: *Proceedings of the 19th ACM SIGKDD International Conference on Knowledge Discovery and Data Mining* KDD '13 pp. 1025–1033, New York, NY, USA: ACM.

Index

www.ingramcontent.com/pod-product-compliance
Lightning Source LLC
Chambersburg PA
CBHW071201210326
41597CB00016B/1633
9784991044526